Distributed Power Generation Scheduling, Modelling and Expansion Planning

Distributed Power Generation Scheduling, Modelling and Expansion Planning

Editors

Javier Contreras
Gregorio Muñoz-Delgado

MDPI • Basel • Beijing • Wuhan • Barcelona • Belgrade • Manchester • Tokyo • Cluj • Tianjin

Editors
Javier Contreras
Universidad de Castilla—La Mancha
Spain

Gregorio Muñoz-Delgado
Universidad de Castilla—La Mancha
Spain

Editorial Office
MDPI
St. Alban-Anlage 66
4052 Basel, Switzerland

This is a reprint of articles from the Special Issue published online in the open access journal *Energies* (ISSN 1996-1073) (available at: https://www.mdpi.com/journal/energies/special_issues/ Distributed_Power_Generation_Scheduling_Modelling_and_Expansion_Planning).

For citation purposes, cite each article independently as indicated on the article page online and as indicated below:

LastName, A.A.; LastName, B.B.; LastName, C.C. Article Title. *Journal Name* **Year**, *Volume Number*, Page Range.

ISBN 978-3-0365-0742-2 (Hbk)
ISBN 978-3-0365-0743-9 (PDF)

Contents

About the Editors

Javier Contreras (Full Professor) received his B.S. degree in Electrical Engineering from the University of Zaragoza, Zaragoza, Spain, in 1989; M.Sc. degree from the University of Southern California, Los Angeles, CA, USA, in 1992; and Ph.D. degree from the University of California, Berkeley, CA, USA, in 1997. He is currently Professor at the Universidad de Castilla-La Mancha, Ciudad Real, Spain. His research interests include power systems planning, operations, and economics, as well as electricity markets. He is an IEEE Fellow.

Gregorio Muñoz-Delgado (Associate Professor) received the Ingeniero Industrial degree, M.Sc. degree, and Ph.D. degree from the Universidad de Castilla-La Mancha, Ciudad Real, Spain, in 2012, 2013, and 2017, respectively. He is currently Associate Professor at the Universidad de Castilla-La Mancha, Ciudad Real, Spain. His research interests are in the fields of power systems planning, operations, and economics. He is a Member of the IEEE Power and Energy Society.

Preface to "Distributed Power Generation Scheduling, Modelling and Expansion Planning"

Due to the decentralization of energy production, electrical distribution systems are in a transition process for the incorporation of distributed energy resources such as renewable-based distributed generators and energy storage units with the aim of achieving an efficient, sustainable, and environmentally friendly network. This new context, where distributed energy resources comes into play, calls for changes in the way distribution systems are operated and planned.

The deployment of renewable-based distributed generators in the network offers a wide variety of technological, economic, and environmental benefits. However, the uncertainty associated with the variability of renewable energy sources significantly hinders the operation and planning of such devices. Nevertheless, the effect of uncertainty may be partially compensated for by the installation of energy storage units, which increases the manageability of the grid. Moreover, demand response and dynamic pricing are also an efficient way to mitigate the uncertainty and improve the system efficiency. In addition, the increasing penetration of electric vehicles represents a new source of generation and demand, depending on whether the vehicles act in a V2G or in a G2V mode, which needs to be accounted for by the system operators and planners. In order to better manage such an amount of distributed energy resources, two different concepts have emerged, namely the aggregator, grouping the energy consumption or generation of several consumers, and the microgrid, which is a small-scale power system that can be operated either independently or grid-connected.

Within this new paradigm, new approaches for the operation and planning of distributed power generation are yet to be explored. This Special Issue aims to integrate them in a comprehensive manner to investigate the resulting effects.

Javier Contreras, Gregorio Muñoz-Delgado
Editors

Article

A Benders' Decomposition Approach for Renewable Generation Investment in Distribution Systems

Sergio Montoya-Bueno, Jose Ignacio Muñoz-Hernandez, Javier Contreras * and Luis Baringo

Department of Applied Mechanics and Project Engineering, University of Castilla-La Mancha,
13071 Ciudad Real, Spain; Sergio.Montoya@uclm.es (S.M.-B.); JoseIgnacio.Munoz@uclm.es (J.I.M.-H.);
Luis.Baringo@uclm.es (L.B.)
* Correspondence: Javier.Contreras@uclm.es

Received: 30 January 2020; Accepted: 21 February 2020; Published: 6 March 2020

Abstract: A model suitable to obtain where and when renewable energy sources (RES) should be allocated as part of generation planning in distribution systems is formulated. The proposed model starts from an existing two-stage stochastic mixed-integer linear programming (MILP) problem including investment and scenario-dependent operation decisions. The aim is to minimize photovoltaic and wind investment costs, operation costs, as well as total substation costs including the cost of the energy bought from substations and energy losses. A new Benders' decomposition framework is used to decouple the problem between investment and operation decisions, where the latter can be further decomposed into a set of smaller problems per scenario and planning period. The model is applied to a 34-bus system and a comparison with a MILP model is presented to show the advantages of the model proposed.

Keywords: Benders' decomposition; distributed generation planning (DGP); two-stage stochastic mixed-integer linear programming (MILP); renewable energy sources (RES)

---

1. Introduction

Distributed generation (DG) has been used to produce energy in remote and isolated places, where the distance between the demand and the producer is short [1]. At present, this trend is changing as it has been proven that DG provides technical, economic, and environmental improvements [2]. The benefits of using DG are based on the reduction of network losses, the voltage level improvement, or the dependence reduction of energy, fuel prices, and traditional generation. All these advantages cause CO_2 emission reduction [3,4]. On the other hand, DG has also disadvantages, mainly related to technical aspects due to the fact that the existing networks have not been designed to incorporate this type of generation [2]. Some drawbacks are reverse current flows, the need for network redesign, or frequency instability [5].

In addition, planning of distribution networks must account for renewable generation uncertainty to meet the future demand in any possible future scenario. In that case, the decisions to be made could be to invest in the network, in substations, in DG generation, or in any combination of them [6]. When the chosen option is DG, it is necessary to determine the type of location, by solving the Distribution Generation Planning (DGP) problem. An example of this sort of problems can be seen in [7], where a linear model that optimizes size and location of DG is used. The objective function maximizes DG real power. In [6], a model is proposed for minimizing the investment and operation DG costs, the cost of the electricity bought from the substation, and the cost of network losses of a distribution company. The model is formulated as a mixed-integer-nonlinear one. Other authors have worked with particle swarm optimization methods [8], where a multi-objective model considers investment and operating costs of new generation, the cost of the energy purchased, and CO_2 emission cost.

A number of references using Benders' decomposition to solve computational-complex problems in distribution [9–12] and transmission systems [13–17] can be found in the technical literature.

The DGP is addressed by a multi-objective mixed-integer linear problem in [9], using Benders' decomposition with an implicit enumeration algorithm where cost and reliability, among others, are included in the objectives' set. No renewable technology is modeled, just feeders and substations. Another multi-objective operation approach is developed in [10]. In this case, the aim is to minimize the total operational costs and emissions, as well as to generate Pareto-optimal solutions for the energy and reserve scheduling problem, using fuzzy decision-making processes. The scenario combination merges wind generation and forecasted demand. The problem is formulated as mixed-integer non-linear problem. Reference [11] studies the point of view of a local distribution company to maximize its profit, using nodal hourly prices within a smart grid. DG, including fossil fuel and renewable units, is taken into account. Reference [12] deals with the day-ahead unit commitment problem in a microgrid system. The problem is formulated as a stochastic mixed integer program that takes into account the uncertainty in PV generation.

A generation expansion planning model is proposed in [13] where the network is disregarded. It includes a generic model for renewable units (wind and solar) and hydro units modeled as storage units. Another generation expansion planning in generalized networks is studied in [14] considering production costs and system reliability in the lower level and the expansion plan in the upper level. A transmission expansion planning model is proposed in [15] incorporating the uncertainty of wind units via scenarios, including the cost of the added lines, and wind curtailment and using a DC power flow. Studies [16,17] propose complementarity models to determine the optimal investment decisions of a profit-oriented private investor interested in building new conventional and wind-power generating units, respectively.

The basis of this paper is a two-stage stochastic mixed-integer linear programming problem (MILP) that is used to determine where and when renewable energy sources should be allocated as part of generation planning in distribution systems [18]. The model proposed in [18] has an important limitation; namely, its computational burden is very high if a large number of scenarios and planning periods is considered. In the worst case, the problem may be even intractable.

Despite this relevant issue, the problem described in [18] has an interesting property: if investment variables are fixed, the problem can be decomposed per scenario and planning period. The proposed Benders' decomposition algorithm takes advantage of this decomposable problem structure to reduce the computational burden of the problem. This is the main benefit of the proposed approach.

Note that, to the best of our knowledge, there is no reference in the technical literature that considers a Benders' decomposition approach for a stochastic multi-stage DGP problem such as the one considered in this paper. The existing literature has not taken into account the benefits of Benders' decomposition for solve the DGP problem with a large number of scenarios of renewable energy. This problem is generally intractable for realistic case studies since it is necessary to consider a large number of scenarios and time periods to obtain informed expansion decisions. Moreover, it includes binary variables that further complicate the problem. Therefore, the traditional resolution methods, as a MILP, for stochastic multi-stage DGP problems are not viable. The Benders' decomposition model presented, although it has greater computational complexity, dramatically reduces resolution times. This model can solve problems with a large number of scenarios and planning periods. In addition, it is able to solve problems that are typically intractable with traditional methods.

Given the above, the contributions of this work are:

1. The stochastic multi-stage DGP problem is formulated using Benders' decomposition that decomposes the problem per both scenario and planning period and
2. A comparison with a standard MILP model is provided and results with a 34-bus test case are shown, where the significant computational advantage of using Benders' decomposition with respect to the MILP model is also shown.

2. Notation

Subscripts t, k, and ω below refer to the values in year t, time block k, and scenario ω, respectively. Superindex (υ) indicates the values in the υ-th Benders' iteration. All indices, sets, constants, parameters and variables used in the document are shown in Table 1.

Table 1. Notation.

Indices and Sets	
Ω^K	Set of indices of time blocks.
Ω^L	Set of indices of load buses.
Ω^N	Set of indices of branches.
Ω^R	Set of indices of blocks used in the piecewise linearization.
Ω^{SS}	Set of indices of substation buses.
Ω^T	Set of indices of years.
Ψ^ω_k	Set of indices of scenarios for the k-th block.
k	Index of time blocks.
n, m	Indices of buses.
r	Index used in the linearization.
t	Index of years.
ω	Index of scenarios.
Constants and Parameters	
ca^{pv}	PV module annualized investment cost [€].
ca^{wd}	Wind turbine annualized investment cost [€].
ca^{SS}	Transformer annualized investment cost [€].
ci^{pv}	PV module investment cost [€].
ci^{wd}	Wind turbine investment cost [€].
ci^{SS}	Transformer investment cost [€].
ci^{bgt}	Investment budget per year [€].
ci^{bgt}_{LC}	Investment budget throughout the life cycle of the new devices [€].
com^{pv}	PV module operation and maintenance costs [€/MWh].
com^{wd}	Wind turbine operation and maintenance costs [€/MWh].
c^{loss}	Loss cost [€/MWh].
c^{ns}	Cost of energy not supplied [€/MWh].
$C^{pv,n}$	Vector of candidate buses n to install PV modules.
$c^{SS}_{k,\omega}$	Cost of the energy purchased by the substation [€/MWh].
$C^{wd,n}$	Vector of candidate buses n to install wind turbines.
d	Discount rate.
f_t	Increasing load factor.
$f^{ld}_{k,\omega}$	Demand factor.
f^{SS}_t	Increasing energy cost factor.
$f^{wd}_{k,\omega}$	Wind turbine power factor available.
$f^{pv}_{k,\omega}$	PV module power factor available.
$\overline{I}^{n,m}$	Maximum current through branch n, m [A_{pu}].
i	Interest rate.
LC^{SS}	Transformer life cycle [years].
LC^{pv}	PV module life cycle [years].
LC^{wd}	Wind turbine life cycle [years].
$m^{n,m,r}_{t,k,\omega}$	Slope of the r-th block of the piecewise linearization for branch n, m.
N^h_k	Number of hours in time block k [hours].

Table 1. *Cont.*

Indices and Sets	
$P^{ld,n}$	Active power load in bus n [MW$_{pu}$].
$\overline{P^{pv}}$	Maximum active power output of PV modules [MW$_{pu}$].
$\overline{P^{wd}}$	Maximum active power output of wind turbine [MW$_{pu}$].
$\overline{P^{node}}$	Maximum active power that can be installed in each bus [MW$_{pu}$].
$Q^{ld,n}$	Reactive power load in bus n [MVar$_{pu}$].
$R^{n,m}$	Resistance of branch n,m [Ω_{pu}].
$\overline{R}$	Number of blocks used in the piecewise linearization.
$\overline{S^{SS}}$	Maximum power output of new transformers [MVA$_{pu}$].
$\overline{S^{NEW,n}}$	Maximum new power allowed for installment in the substation in bus n [MVA$_{pu}$].
$S^{SS,n}$	Existing power in the substation in bus n [MVA$_{pu}$].
S^{base}	Power base [MVA].
$tan\left(\varphi^{SS}\right)$	Tangent angle in substation.
$tan\left(\varphi^{pv}\right)$	Tangent angle of PV modules.
$tan\left(\varphi^{wd}\right)$	Tangent angle of wind turbines.
$\underline{V}$	Minimum voltage magnitude [kV$_{pu}$].
$\overline{V}$	Maximum voltage magnitude [kV$_{pu}$].
V^{nom}	Nominal voltage of the distribution network [kV$_{pu}$].
$X^{n,m}$	Reactance of branch n,m [Ω_{pu}].
$\overline{Y^{pv,n}}$	Maximum number of PV modules to be installed in bus n.
$\overline{Y^{wd,n}}$	Maximum number of wind turbines to be installed in bus n.
$Z^{n,m}$	Impedance of branch n,m [Ω_{pu}].
$\gamma_{k,\omega}$	Weight of scenario ω in time block k.
$\Delta S^{n,m,r}_{t,k,\omega}$	Upper bound of each r block of the power flow through branch n,m [MVA$_{pu}$].
β_t	Present worth factor.

Variables	
ci_t	Investment cost [€].
$closs_{t,k,\omega}$	Losses cost [€/h].
$cns_{t,k,\omega}$	Penalty for not supplied energy [€/h].
$cnew_{t,k,\omega}$	Maintenance and operation costs of DG candidates [€/h].
$com_{t,k,\omega}$	Maintenance and operation total costs [€/h].
$css_{t,k,\omega}$	Cost of energy purchased by the substation [€/h].
$I^{sqr,n,m}_{t,k,\omega}$	Square of current flow magnitude of branch n,m [A^2_{pu}].
$PT^{wd,n}_t$	Active power of wind turbines to be installed in bus n [MW$_{pu}$].
$PT^{pv,n}_t$	Active power of PV modules to be installed in bus n [MW$_{pu}$].
$P^{ns,n}_{t,k,\omega}$	Not supplied active power in bus n [MW$_{pu}$].
$P^{pv,n}_{t,k,\omega}$	Active power injected by PV modules in bus n [MW$_{pu}$].
$P^{wd,n}_{t,k,\omega}$	Active power injected by wind turbines in bus n [MW$_{pu}$].
$P^{SS,n}_{t,k,\omega}$	Active power purchased by the substation in bus n [MW$_{pu}$].
$P^{+,n,m}_{t,k,\omega}$	Active power flow in branch n,m in the forward direction [MW$_{pu}$].
$P^{-,n,m}_{t,k,\omega}$	Active power flow in branch n,m in the backward direction [MW$_{pu}$].
$Q^{ns,n}_{t,k,\omega}$	Not supplied reactive power in bus n [MVAr$_{pu}$].
$Q^{pv,n}_{n,t,k,\omega}$	Reactive power injected by PV modules in bus n [MVAr$_{pu}$].
$Q^{wd,n}_{t,k,\omega}$	Reactive power injected by wind turbines in bus [MVAr$_{pu}$].
$Q^{SS,n}_{t,k,\omega}$	Reactive power purchased by the substation in bus n [MVAr$_{pu}$].
$Q^{+,n,m}_{t,k,\omega}$	Reactive power flow through branch n,m in the forward direction [MVAr$_{pu}$].
$Q^{-,n,m}_{t,k,\omega}$	Reactive power flow through branch n,m in the backward direction [MVAr$_{pu}$].
$ST^{SS,n}_t$	Total available power in the substation in bus n [MVA$_{pu}$].
$S^{NEW;n}_t$	New power installed in the substation in bus n [MVA$_{pu}$].

Table 1. *Cont.*

Indices and Sets	
$V_{t,k,w}^{sqr,n}$	Square of voltage magnitude in node n $[kV_{pu}^2]$.
$Y_t^{pv,n}$	Number of PV modules to be installed in bus n.
$Y_t^{wd,n}$	Number of wind turbines to be installed in bus n.
$Y_t^{SS,n}$	Number of transformers to be installed in bus n.
$Y_{t,k,\omega}^{P+,n,m}$	Binary variable that defines if the active power flow through branch n,m is in the forward direction.
$Y_{t,k,\omega}^{P-,n,m}$	Binary variable that defines if the active power flow through branch n,m is in the backward direction.
$Y_{t,k,\omega}^{Q+,n,m}$	Binary variable that defines if the reactive power flow through branch n,m is in the forward direction.
$Y_{t,k,\omega}^{Q-,n,m}$	Binary variable that defines if the reactive power flow through branch n,m is in the backward direction.
$\Delta P_{t,k,\omega}^{n,m,r}$	Value of the r-th block associated with the active power through branch n,m $[MW_{pu}]$.
$\Delta Q_{t,k,\omega}^{n,m,r}$	Value of the r-th block associated with the reactive power through branch n,m $[MVAr_{pu}]$.

3. Problem Formulation

3.1. Objective Function

The goal of the model is to minimize the total system costs (TSC) considering DG. The model utilizes a two-stage stochastic mixed-integer linear programming model. The investment variables that do not depend on the scenarios are established in the first stage. In the second stage, dependent or stochastic operation variables that depend on the scenarios are determined.

TSC are composed of two terms (Equation (1)). The first term corresponds to the first-stage variables that determine the number of new units such as wind turbines, PV modules, and transformers to install. The second term corresponds to the second-stage variables whose values are obtained after the outcome of scenario ω is known. See [18] for details.

$$\min TSC = \sum_{t \in \Omega^T} \beta_t \left(ci_t + \sum_{k \in \Omega^K} N_k^h \sum_{\omega \in \Psi_k^\omega} \gamma_{k,\omega} com_{t,k,\omega} \right). \tag{1}$$

Total costs are updated using the present worth factor, $\beta_t = 1/(1+d)^t$.

The interest rate, i, is used to calculate the annualized investment cost payment rates. The payment contributions for the three technologies, namely, transformers, wind turbines, and PV modules, are obtained in Equations (2), (3) and (4), respectively.

$$ca^{SS} = \frac{ci^{SS}(1+i)^{LC^{SS}}}{(1+i)^{LC^{SS}} - 1} \tag{2}$$

$$ca^{pv} = \frac{ci^{pv}(1+i)^{LC^{pv}}}{(1+i)^{LC^{pv}} - 1} \tag{3}$$

$$ca^{wd} = \frac{ci^{wd}(1+i)^{LC^{wd}}}{(1+i)^{LC^{wd}} - 1}. \tag{4}$$

After having determined the payment rates, the investment costs are calculated as shown in Equations (5) and (6).

$$ci_t = \sum_{n \in \Omega^{SS}} ca^{SS} Y_t^{SS,n} + \sum_{n \in \Omega^L} \left(ca^{pv} Y_t^{pv,n} + ca^{wd} Y_t^{wd,n} \right); t = 1 \tag{5}$$

$$ci_t = \sum_{n\in\Omega^{SS}} ca^{SS} Y_t^{SS,n} + \sum_{n\in\Omega^L} \left(ca^{pv} Y_t^{pv,n} + ca^{wd} Y_t^{wd,n} \right) + ci_{t-1}; \; t > 1. \tag{6}$$

The operation and maintenance total costs (Equation (7)) take into account the cost of losses (Equation (8)), the penalty for the energy not supplied (Equation (9)), the cost of purchase of energy from substations (Equation (10)), as well as renewable DG candidates' operation and maintenance costs (Equation (11)).

$$com_{t,k,\omega} = closs_{t,k,\omega} + cns_{t,k,\omega} + css_{t,k,\omega} + cnew_{t,k,\omega}; \; \forall (t, k, \omega) \tag{7}$$

$$closs_{t,k,\omega} = c^{loss} \sum_{n,m\in\Omega^N} S^{base} R^{n,m} I_{t,k,\omega}^{sqr,n,m}; \; \forall (t, k, \omega) \tag{8}$$

$$cns_{t,k,\omega} = c^{ns} \sum_{n\in\Omega^L} S^{base} P_{t,k,\omega}^{ns,n}; \; \forall (t, k, \omega) \tag{9}$$

$$css_{t,k,\omega} = c_{k,\omega}^{SS} f_t^{SS} \sum_{n\in\Omega^{SS}} S^{base} P_{t,k,\omega}^{SS,n}; \; \forall (t, k, \omega) \tag{10}$$

$$cnew_{t,k,\omega} = S^{base} \sum_{n\in\Omega^L} \left(com^{pv} P_{t,k,\omega}^{pv,n} + com^{wd} P_{t,k,\omega}^{wd,n} \right); \; \forall (t, k, \omega). \tag{11}$$

3.2. Constraints

The investments in new transformers (Equation (12)), wind units (Equation (13)), and PV modules (Equation (14)) are limited per node n.

$$0 \le \sum_{t\in\Omega^T} Y_t^{SS;n} \le \overline{S^{NEW;n}} / \overline{S^{SS}}; \; \forall n \in \Omega^{SS} \tag{12}$$

$$0 \le \sum_{t\in\Omega^T} Y_t^{wd,n} \le \overline{Y_n^{wd}}; \; \forall n \in \Omega^L \tag{13}$$

$$0 \le \sum_{t\in\Omega^T} Y_t^{pv,n} \le \overline{Y_n^{pv}}; \; \forall n \in \Omega^L. \tag{14}$$

The maximum value of the power allowed to install at each bus n is limited by Equation (15).

$$\overline{Pnode} \ge \sum_{t\in\Omega^T} \left(\overline{Pwd} \; Y_t^{wd,n} + \overline{Ppv} \; Y_t^{pv,n} \right); \; \forall n \in \Omega^L. \tag{15}$$

The investments are limited by Equations (16) and (17). Equation (16) refers to the annual investment payment limit that describes the budget available for each investment period, while the portfolio investment (Equation (17)) is related to the amount of money that is available for investment in the long term.

$$ci_t \le ci^{bgt}; \; \forall t. \tag{16}$$

$$\sum_{t\in\Omega^T} \beta^t \left[\sum_{n\in\Omega^{SS}} ci^{SS} Y_t^{SS,n} + \sum_{n\in\Omega^L} \left(ci^{pv} Y_t^{pv,n} + ci^{wd} Y_t^{wd,n} \right) \right] \le ci_{LC}^{bgt}. \tag{17}$$

The power flow constraints (Equations (18)–(41)) refer to both active and reactive power equations. The distribution system is composed of two types of buses (load and substation buses) where each kind of bus has a different load flow expression.

$$\sum_{n\in\Omega^N} \left(P_{t,k,\omega}^{+,n,m} - P_{t,k,\omega}^{-,n,m} \right) + P_{t,k,\omega}^{ns,m} + P_{t,k,\omega}^{wd,m} + P_{t,k,\omega}^{pv,m} - \sum_{n\in\Omega^N} \left(P_{t,k,\omega}^{+,m,n} - P_{t,k,\omega}^{-,m,n} + R^{m,n} I_{t,k,\omega}^{sqr,m,n} \right)$$
$$= f_t f_{k,\omega}^{ld} P^{ld,m}; \; \forall \left(m \in \Omega^L, t, k, \omega \right) \tag{18}$$

$$\sum_{n \in \Omega^N} \left(P_{t,k,\omega}^{+,n,m} - P_{t,k,\omega}^{-,n,m}\right) + P_{t,k,\omega}^{SS,m} - \left(P_{t,k,\omega}^{+,m,n} - P_{t,k,\omega}^{-,m,n} + R^{m,n}\, I_{t,k,\omega}^{sqr,m,n}\right) = f_t f_{k,\omega}^{fld} P^{ld,m}; \ \forall\left(m \in \Omega^{SS}, t, k, \omega\right) \quad (19)$$

$$\sum_{n \in \Omega^N} \left(Q_{t,k,\omega}^{+,n,m} - Q_{t,k,\omega}^{-,n,m}\right) + Q_{t,k,\omega}^{ns,m} + Q_{t,k,\omega}^{wd,m} + Q_{t,k,\omega}^{pv,m} - \left(Q_{t,k,\omega}^{+,m,n} - Q_{t,k,\omega}^{-,m,n} + X^{m,n}\, I_{t,k,\omega}^{sqr,m,n}\right)$$
$$= f_t f_{k,\omega}^{fld} Q^{ld,m} 0; \ \forall\left(m \in \Omega^{L}, t, k, \omega\right) \quad (20)$$

$$\sum_{n \in \Omega^N} \left(Q_{t,k,\omega}^{+,n,m} - Q_{t,k,\omega}^{-,n,m}\right) + Q_{t,k,\omega}^{SS,m} - \left(Q_{t,k,\omega}^{+,m,n} - Q_{t,k,\omega}^{-,m,n} + X^{m,n}\, I_{t,k,\omega}^{sqr,m,n}\right)$$
$$= f_t f_{k,\omega}^{fld} Q^{ld,m}; \ \forall\left(m \in \Omega^{SS}, t, k, \omega\right). \quad (21)$$

The voltage is related to the different electrical values (Equation (22)) and is limited by the maximum and minimum values (Equation (23)).

$$V_{t,k,w}^{sqr,m} - Z^{m,n2} I_{t,k,\omega}^{sqr,m,n} - V_{t,k,w}^{sqr,n} - 2\left(R^{m,n}\left(P_{t,k,\omega}^{+,m,n} - P_{t,k,\omega}^{-,m,n}\right) + X^{m,n}\left(Q_{t,k,\omega}^{+,m,n} - Q_{t,k,\omega}^{-,m,n}\right)\right)$$
$$= 0; \ \forall\left(m, n \in \Omega^N, t, k, \omega\right) \quad (22)$$

$$\underline{V}^2 \le V_{t,k,w}^{sqr,m} \le \overline{V}^2; \ \forall\left(m \in \Omega^N, t, k, \omega\right). \quad (23)$$

Equation (24) represents the maximum current permitted through the distribution feeders. These limits are applied to the active (25) and (26) and reactive power Equations (27) and (28). Equations (29) and (30) are necessary to avoid inverse flows.

$$0 \le I_{t,k,\omega}^{sqr,m,n} \le \overline{I^{m,n}}^2; \ \forall\left(m, n \in \Omega^N, t, k,' \omega\right) \quad (24)$$

$$P_{t,k,\omega}^{+,m,n} \le V^{nom}\, \overline{I^{m,n}}\, Y_{t,k,\omega}^{P+,m,n}; \ \forall\left(m, n \in \Omega^N, t, k, \omega\right) \quad (25)$$

$$P_{t,k,\omega}^{-,m,n} \le V^{nom}\, \overline{I^{m,n}}\, Y_{t,k,\omega}^{P-,m,n}; \ \forall\left(m, n \in \Omega^N, t, k, \omega\right) \quad (26)$$

$$Q_{t,k,\omega}^{+,m,n} \le V^{nom}\, \overline{I^{m,n}}\, Y_{t,k,\omega}^{Q+,m,n}; \ \forall\left(m, n \in \Omega^N, t, k, \omega\right) \quad (27)$$

$$Q_{t,k,\omega}^{-,m,n} \le V^{nom}\, \overline{I^{m,n}}\, Y_{t,k,\omega}^{Q-,m,n}; \ \forall\left(m, n \in \Omega^N, t, k, \omega\right) \quad (28)$$

$$Y_{t,k,\omega}^{P+,m,n} + Y_{t,k,\omega}^{P-,m,n} \le 1; \ \forall\left(m, n \in \Omega^N, t, k, \omega\right) \quad (29)$$

$$Y_{t,k,\omega}^{Q+,m,n} + Y_{t,k,\omega}^{Q-,m,n} \le 1; \ \forall\left(m, n \in \Omega^N, t, k, \omega\right) \quad (30)$$

The power flow has been linearized using Equations (31) to (41) as explained in [18] and [19].

$$Y_{t,k,\omega}^{P+;n,m} \in \{0,1\}; \ \forall\left(m, n \in \Omega^N, t, k, \omega\right) \quad (31)$$

$$Y_{t,k,\omega}^{P-;n,m} \in \{0,1\}; \ \forall\left(m, n \in \Omega^N, t, k, \omega\right) \quad (32)$$

$$Y_{t,k,\omega}^{Q+;n,m} \in \{0,1\}; \ \forall\left(m, n \in \Omega^N, t, k, \omega\right) \quad (33)$$

$$Y_{t,k,\omega}^{Q-;n,m} \in \{0,1\}; \ \forall\left(m, n \in \Omega^N, t, k, \omega\right) \quad (34)$$

$$0 \le \Delta P_{t,k,\omega}^{n,m,r} \le \Delta S_{t,k,\omega}^{m,n,r}; \ \forall\left(m, n \in \Omega^N, r \in \Omega^R, t, k, \omega\right) \quad (35)$$

$$0 \le \Delta Q_{t,k,\omega}^{n,m,r} \le \Delta S_{t,k,\omega}^{m,n,r}; \ \forall\left(m, n \in \Omega^N, r \in \Omega^R, t, k, \omega\right) \quad (36)$$

$$\sum_{r \in \Omega^R} \left(m_{t,k,\omega}^{m,n,r}\, \Delta P_{t,k,\omega}^{m,n,r}\right) + \sum_{r \in \Omega^R} \left(m_{t,k,\omega}^{m,n,r}\, \Delta Q_{t,k,\omega}^{m,n,r}\right) = V^{nom2} I_{t,k,\omega}^{sqr;m,n}; \ \forall\left(m, n \in \Omega^N, t, k, \omega\right) \quad (37)$$

$$P_{t,k,\omega}^{+;m,n} + P_{t,k,\omega}^{-;m,n} = \sum_{r \in \Omega^R} \Delta P_{t,k,\omega}^{m,n,r}; \ \forall\left(m, n \in \Omega^N, t, k, \omega\right) \quad (38)$$

$$Q_{t,k,\omega}^{+;m,n} + Q_{t,k,\omega}^{-;m,n} = \sum_{r \in \Omega^R} \Delta Q_{t,k,\omega}^{m,n,r}; \ \forall\left(m, n \in \Omega^N, t, k, \omega\right) \tag{39}$$

$$m_{m,n,r,t,k,\omega} = (2r - 1)\Delta S_{m,n,r,t,k,\omega}; \ \forall\left(m, n \in \Omega^N, r \in \Omega^R, t, k, \omega\right) \tag{40}$$

$$\Delta S_{m,n,r,t,k,\omega} = \frac{\left(V_{nom} \ \overline{I_{m,n}}\right)}{\overline{R}}; \ \forall\left(m, n \in \Omega^N, r \in \Omega^R, t, k, \omega\right). \tag{41}$$

The not supplied active and reactive power must be less than the active and reactive power demand, respectively. Note that the demand is updated each period.

$$0 \leq P_{t,k,\omega}^{ns,n} \leq f_t f_{k,\omega}^{ld} P^{ld,n}; \ \forall\left(n \in \Omega^L, t, k, \omega\right) \tag{42}$$

$$0 \leq Q_{t,k,\omega}^{ns,n} \leq f_t f_{k,\omega}^{ld} Q^{ld,n}; \ \forall\left(n \in \Omega^L, t, k, \omega\right). \tag{43}$$

The substation available power is subject to yearly updates (44). New investments are made considering the already existing power and the new possible power of the candidate transformers (45), (46).

$$ST_t^{SS;n} = S^{SS;n} + S_t^{NEW;n}; \ \forall\left(n \in \Omega^{SS}, t\right) \tag{44}$$

$$S_t^{NEW;n} = Y_t^{SS;n} \ \overline{S^{SS}} / S^{base}; \ \forall\left(n \in \Omega^{SS}, t = 1\right) \tag{45}$$

$$S_t^{NEW;n} = S_{t-1}^{NEW;n} + Y_t^{SS;n} \ \overline{S^{SS}} / S^{base}; \ \forall\left(n \in \Omega^{SS}, t > 1\right). \tag{46}$$

The renewable power (wind turbines and PV modules) available is updated every year. Note that the vector of the candidate nodes appears in Equations (47) to (50).

$$PT_t^{wd,n} = \overline{P^{wd}} \ Y_t^{wd;n} \ C^{wd;n}; \ \forall\left(n \in \Omega^L, t = 1\right) \tag{47}$$

$$PT_t^{wd,n} = \overline{P^{wd}} \ Y_t^{wd,n} \ C^{wd;n} + PT_{t-1}^{wd,n}; \ \forall\left(n \in \Omega^L, t > 1\right) \tag{48}$$

$$PT_t^{pv,n} = \overline{P^{pv}} \ Y_t^{pv,n} \ C^{pv,n}; \ \forall\left(n \in \Omega^L, t = 1\right) \tag{49}$$

$$PT_t^{pv,n} = \overline{P^{pv}} \ Y_t^{pv,n} \ C^{pv,n} + PT_{n,t-1}^{pv,n}; \ \forall\left(n \in \Omega^L, t > 1\right). \tag{50}$$

The maximum amount of generation that is available is limited by the generation levels of each of the technologies available (Equations (51) and (52)).

$$0 \leq P_{t,k,\omega}^{wd,n} \leq f_{k,\omega}^{wd} PT_t^{wd,n}; \ \forall\left(n \in \Omega^L, t, k, \omega\right) \tag{51}$$

$$0 \leq P_{t,k,\omega}^{pv,n} \leq f_{k,\omega}^{pv} PT_t^{pv,n}; \ \forall\left(n \in \Omega^L, t, k, \omega\right). \tag{52}$$

The active power injected into the distribution system must keep a constant power factor (Equation (53)). In this way the reactive power is also limited (Equation (54)).

$$P_{t,k,\omega}^{SS,n} \leq ST_t^{SS,n} / \sqrt{\left(1 + tan(\varphi^{SS})^2\right)}; \ \forall\left(n \in \Omega^{SS}, t, k, \omega\right) \tag{53}$$

$$Q_{t,k,\omega}^{SS,n} \leq tan\left(\varphi^{SS}\right) P_{t,k,\omega}^{SS,n}; \ \forall\left(n \in \Omega^{SS}, t, k, \omega\right). \tag{54}$$

The reactive renewable power that is injected in the distribution system must have a constant power factor (Equations (55) and (56)).

$$0 \leq Q_{t,k,\omega}^{wd,n} \leq P_{t,k,\omega}^{wd,n} \ tan\left(\varphi^{wd}\right); \ \forall\left(n \in \Omega^L, t, k\right) \tag{55}$$

$$0 \leq Q_{t,k,\omega}^{pv,n} \leq P_{t,k,\omega}^{pv,n} \ tan(\varphi^{pv}); \ \forall\left(n \in \Omega^L, t, k\right). \tag{56}$$

4. Solution Procedure: Benders' Decomposition

Model (1)–(56) is set as a MILP problem whose size depends on the number of time blocks, scenarios, and years considered in the study. In order to obtain informed expansion decisions, a large number of time blocks and scenarios are needed to obtain a good representation of the loads and renewable power outputs, as well as a large number of years. However, in such a case the MILP problem (1)–(56) may become computationally intractable. Nevertheless, note that if variables $Y_t^{SS;n}$, $Y_t^{pv;n}$, and $Y_t^{wd;n}$, i.e., the expansion decision variables, are fixed to given values, then, problem (1)–(56) can be decomposed into a set of smaller problems, one for each time block, scenario, and year. This allows us to use Benders' decomposition to solve the problem.

Model (1)–(56) uses binary variables for the linearization of power flow equations. Since Benders' decomposition cannot have binary variables in the sub-problem, a procedure to fix the values of these variables must be used. Further details are shown in step 1.

In the following sections superindex (v) indicates the values in the v-th Benders' iteration.

The proposed Benders' decomposition algorithm comprises the steps below:

1. Step 0. Initialization: Initialize the iteration counter, $v = 1$, the upper bound of the objective function, $z_{up}^{(1)} = \infty$ and its lower bound, $z_{down}^{(1)} = -\infty$ and solve the following problem:

$$\underset{Y_t^{SS;n},\, Y_t^{pv;n},\, Y_t^{wd;n},\, \alpha}{minimize} \quad \sum_{t \in \Omega^T} \beta_t ci_t + \alpha \tag{57}$$

subject to constraints (12)–(17) and

$$\alpha \geq \alpha^{down}. \tag{58}$$

This problem has the trivial solution: $\alpha^{(1)} = \alpha^{down}$., $Y_t^{SS;n(1)}$, $Y_t^{pv;n(1)}$, $Y_t^{wd;n(1)} = 0$, $\forall(t,n)$.

After problem (57), (58), (12)–(17)above is solved, the lower bound of the optimal value of the objective function (1) is updated:

$$z_{down}^{(1)} = \sum_{t \in \Omega^T} \beta_t ci_t^{(1)} + \alpha^{(1)}. \tag{59}$$

2. Step 1. Sub-problem solution: The variables $Y_t^{SS;n}$, $Y_t^{pv;n}$, and $Y_t^{wd;n}$, $\forall(t,n)$ are set, to their optimal values from the previous step and solve the following sub-problems, one for each year t, time block k, and scenario ω.

$$\left\{ \underset{\Xi_{t,k,\omega}^1}{minimize} \quad \beta_t N_k^h \gamma_{k,\omega} com_{t,k,\omega} \right. \tag{60}$$

subject to constraints (7)–(11), (18)–(56) and

$$Y_t^{SS;n} = Y_t^{SS;n(v)} \; ; \forall n \in \Omega^{SS} \tag{61}$$

$$Y_t^{pv;n} = Y_t^{pv;n(v)} \; ; \forall n \in \Omega^L \tag{62}$$

$$\left. Y_t^{wd;n} = Y_t^{wd;n(v)} ; \forall n \in \Omega^L \right\}; \; \forall(t,k,\omega) \tag{63}$$

where:

$$\begin{aligned}
\Xi_{t,k,\omega}^1 = \; & \{ I_{t,k,\omega}^{sqr;n,m}, P_{t,k,\omega}^{ns;n}, P_{t,k,\omega}^{pv;n}, P_{t,k,\omega}^{wd;n}, P_{t,k,\omega}^{SS;n}, P_{t,k,\omega}^{+;n,m}, P_{t,k,\omega}^{-;n,m}, Q_{t,k,\omega}^{ns;n}, Q_{t,k,\omega}^{SS;n}, Q_{n,t,k,\omega}^{pv;n}, Q_{t,k,\omega}^{wd;n}, Q_{t,k,\omega}^{+;n,m}, Q_{t,k,\omega}^{-;n,m}, V_{t,k,w}^{sqr;n}, Y_{t,k,\omega}^{P+;n,m}, \\
& Y_{t,k,\omega}^{P-;n,m}, Y_{t,k,\omega}^{Q+;n,m}, Y_{t,k,\omega}^{Q-;n,m}, \Delta P_{t,k,\omega}^{n,m,r}, \Delta Q_{t,k,\omega}^{n,m,r} \}
\end{aligned}$$

Each instance of the sub-problem, one for each year t, time block k, and scenario ω, is solved twice. The first simulation solves problem (60)–(63), (7)–(11), (18)–(56) as a MILP problem, to obtain the values of $Y_{t,k,\omega}^{P+,m,n}$, $Y_{t,k,\omega}^{P-,m,n}$, $Y_{t,k,\omega}^{Q+,m,n}$, and $Y_{t,k,\omega}^{Q-;n,m}$, which are the sub-problem's binary variables. Then, the second run uses the optimal values of variables $Y_{t,k,\omega}^{P+,m,n}$, $Y_{t,k,\omega}^{P-,m,n}$, $Y_{t,k,\omega}^{Q+,m,n}$, and $Y_{t,k,\omega}^{Q-;n,m}$, obtained in the first simulation, as parameters for solving sub-problem (64)–(67), (7)–(11), (18)–(56) i.e., in this case we solve sub-problem (64)–(67), (7)–(11), (18)–(56) as the following LP problem.

$$\left\{ \begin{array}{l} \textit{minimize} \\ \Xi^2_{t,k,\omega} \end{array} \quad \beta_t N_k^h \, \gamma_{k,\omega} com_{t,k,\omega} \right. \tag{64}$$

subject to constraints (7)–(11), (18)–(56) and

$$Y_t^{SS;n} = Y_t^{SS;n(v)} \;:\; \lambda_{1,n,t,k,\omega}; \; \forall n \in \Omega^{SS} \tag{65}$$

$$Y_t^{pv;n} = Y_t^{pv;n(v)} \;:\; \lambda_{2,n,t,k,\omega}; \; \forall n \in \Omega^L \tag{66}$$

$$Y_t^{wd;n} = Y_t^{wd;n(v)} \;:\; \lambda_{3,n,t,k,\omega}; \; \forall n \in \Omega^L\big\}; \; \forall(t,k,\omega). \tag{67}$$

where:

$$\Xi^2_{t,k,\omega} = \big\{ I_{t,k,\omega}^{sqr;n,m}, P_{t,k,\omega}^{ns;n}, P_{t,k,\omega}^{pv;n}, P_{t,k,\omega}^{wd;n}, P_{t,k,\omega}^{SS;n}, Q_{t,k,\omega}^{ns;n}, Q_{t,k,\omega}^{SS;n}, Q_{n,t,k,\omega}^{pv;n}, Q_{t,k,\omega}^{wd;n}, V_{t,k,w}^{sqr;n}, \Delta P_{t,k,\omega}^{n,m,r}, \Delta Q_{t,k,\omega}^{n,m,r} \big\}.$$

The outputs of sub-problem (60)–(63), (7)–(11), (18)–(56) are variables of set $\Xi^1_{t,k,\omega}$, and the outputs of sub-problem (64)–(67), (7)–(11), (18)–(56) are variables of set $\Xi^2_{t,k,\omega}$, as well as dual variables $\lambda_{1,n,t,k,\omega}$, $\lambda_{2,n,t,k,\omega}$, and $\lambda_{3,n,t,k,\omega}$. Note that the sensitivities used for Benders' cuts in (65), (66) and (67) are only obtained after the second run.

A sub-problem (64)–(67), (7)–(11), (18)–(56) is solved for each year t, time block k, and scenario ω. In order to formulate the Benders' cuts in the Master problem (see Step 3), it is necessary to define and compute parameters $\tilde{\lambda}_{1,n,t}$, $\tilde{\lambda}_{2,n,t}$, and $\tilde{\lambda}_{3,n,t}$ as follows:

$$\tilde{\lambda}_{1,n,t} = \sum_{k\in\Omega^K} \sum_{\omega\in\Psi_k^\omega} \lambda_{1,n,t,k,\omega}; \; \forall\big(n \in \Omega^{SS}, t\big) \tag{68}$$

$$\tilde{\lambda}_{2,n,t} = \sum_{k\in\Omega^K} \sum_{\omega\in\Psi_k^\omega} \lambda_{2,n,t,k,\omega}; \; \forall\big(n \in \Omega^L, t\big) \tag{69}$$

$$\tilde{\lambda}_{3,n,t} = \sum_{k\in\Omega^K} \sum_{\omega\in\Psi_k^\omega} \lambda_{3,n,t,k,\omega}; \; \forall\big(n \in \Omega^L, t\big). \tag{70}$$

Finally, the upper bound of the optimal value of the objective function (1) is updated as:

$$z_{up}^{(v)} = \sum_{t\in\Omega^T} \beta_t ci_t^{(v)} + \sum_{t\in\Omega^T} \sum_{k\in\Omega^K} \sum_{\omega\in\Psi_k^\omega} \beta_t N_k^h \, \gamma_{k,\omega} com_{t,k,\omega}^{(v)}. \tag{71}$$

3. Step 2: Convergence checking: Check if the difference between the upper and the lower bounds, $z_{up}^{(v)} - z_{down}^{(v)}$ is lower than a predefined tolerance, ε. If so, the algorithm has converged, and the optimal solution consists of variables $Y_t^{SS;n}$, $Y_t^{pv;n}$, $Y_t^{wd;n}$, $\forall(t,n)$; as well as the variables in set Ξ for iteration (v). If not, the algorithm proceeds to the next step.

4. Step 3: Master problem solution: Update the iteration counter, $v \leftarrow v + 1$, and solve the following master problem:

$$\begin{array}{l} \textit{minimize} \\ Y_t^{SS;n}, Y_t^{pv;n}, Y_t^{wd;n}, \alpha \end{array} \quad \sum_{t\in\Omega^T} \beta_t ci_t + \alpha \tag{72}$$

subject to constraints (12)–(17) and

$$\alpha \geq \alpha^{down} \tag{73}$$

$$\sum_{t,k,\omega} \beta_t N_k^h \, \gamma_{k,\omega} com_{t,k,\omega}{}^{(v)}$$

$$+ \sum_{t\in\Omega^T}\left(\sum_{n\in\Omega^{SS}} \widetilde{\lambda}_{1,n,t}^{(h)}\left(Y_t^{SS;n} - Y_t^{SS;n(v)} \right)\right.$$
$$\left. + \sum_{n\in\Omega^L}\left[\widetilde{\lambda}_{2,n,t}^{(h)}\left(Y_t^{pv;n} - Y_t^{pv;n(v)} \right) + \widetilde{\lambda}_{3,n,t}^{(h)}\left(Y_t^{wd;n} - Y_t^{wd;n(h)} \right)\right]\right) \leq \alpha; \; \forall h \tag{74}$$
$$= 1,\dots v-1$$

where ci_t is defined in Equations (5) and (6) and their parameters in Equations (2)–(4).

Each constraint in (74) is known as a Benders' cut [20]. Benders' cuts approximate objective function (1) from below as a function of the investment variables. Notice that at every iteration the size of master problem (72)–(74), (12)–(17) increases since a new constraint is added.

Next, the lower bound of the objective function is updated (1):

$$z_{down}^{(v)} = \sum_{t\in\Omega^T} \beta_t ci_t{}^{(v)} + \alpha^{(v)}. \tag{75}$$

Then, the algorithm continues at Step 1.

For clarity purposes, the flowchart of the algorithm is shown in Figure 1.

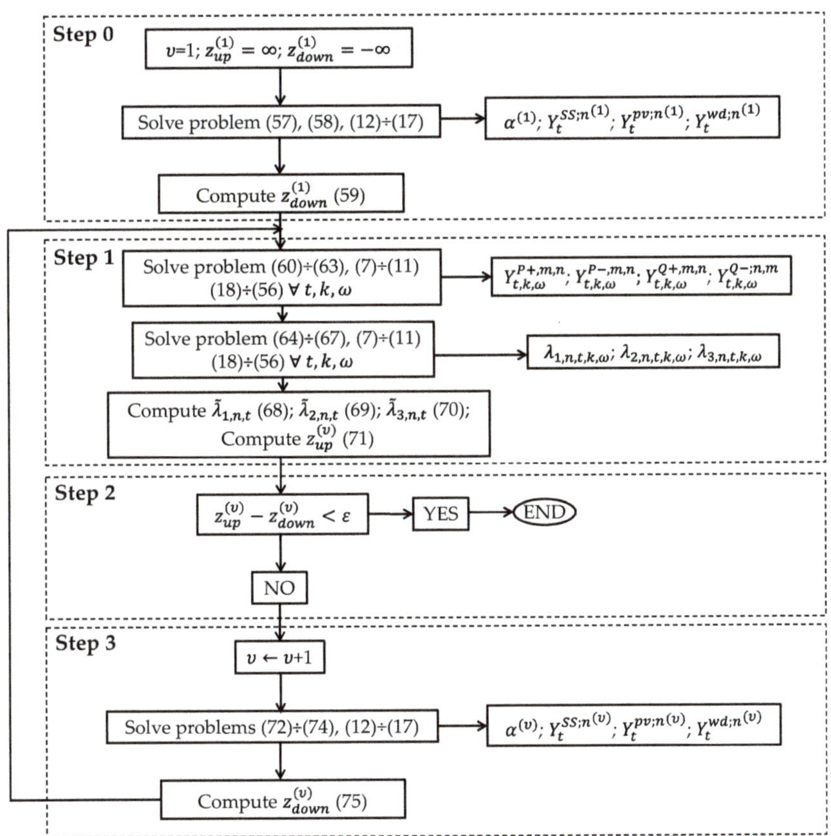

Figure 1. Benders' decomposition flowchart.

Note that the Benders' decomposition algorithm described in this section can solve each sub-problem separately, and hence, the computational time of the proposed algorithm grows linearly with the number of scenarios, time blocks, and years, being its scalability excellent.

5. Case Study Data

This section describes the data used for the simulations of the problem explained in Section 3. The Benders' decomposition method is applied to a 34-bus three-phase radial feeder [21], which has 1 substation, 29 buses with load, and 5 buses with no load. The topology of the system is shown in Figure 2. The data are based on those provided in [18] and [21], but adding new scenarios. In [18] there are 3 levels of demand, wind production, and solar production per each of the 8 time blocks. Now, there are 4 levels and 12 time blocks.

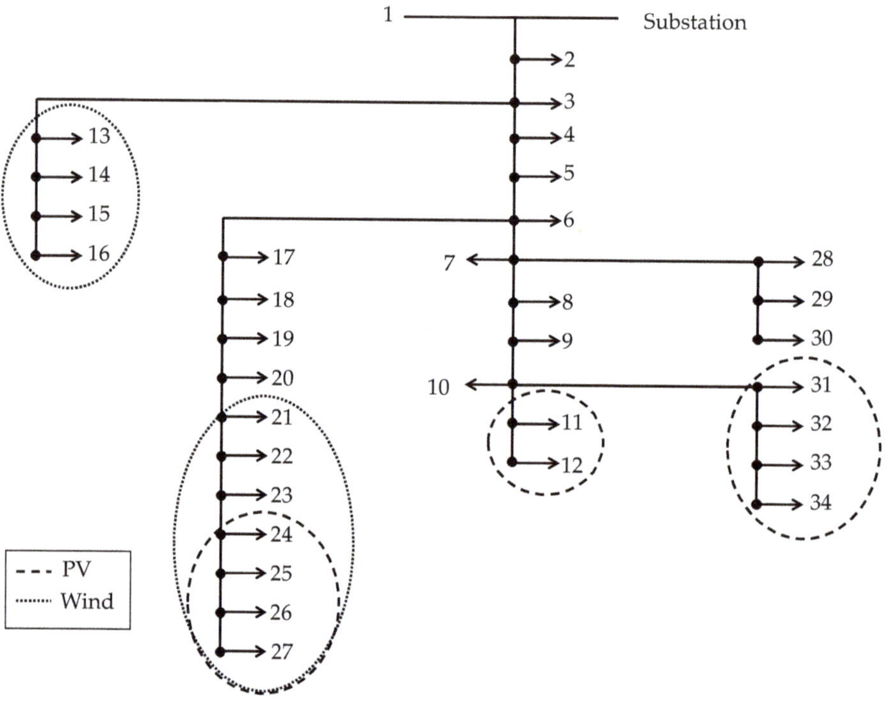

Figure 2. Distribution system under study.

The wind turbines chosen have a capacity of 100 kW and the PV modules have power of 2.5 kW. The candidate buses for each technology are shown in Figure 2 and the maximum power (wind turbine and PV modules) that may be installed at each bus is 250 kW. The maximum number of units is limited to 2 for wind turbines and to 85 for PV modules, and the substation can be expanded adding transformers of sizes ranging from 1 MVA up to a maximum of 5 MVA. Demand increases 2% each year along the planning horizon (20 years). The voltage data of the distribution system is 1 pu in the substation node and the minimum and maximum allowable voltage values are 0.95 pu and 1.05 pu, respectively. The values defined for the interest and discount rates are 8% and 12.5% [6], respectively. Investment data of new devices are shown in Table 2.

Table 2. Investment data.

Unit	Investment Cost [€]	Life Cycle [years]	Tan (φ)
Transformer	20,000	20	0.48
Wind turbines	125,155	20	0.48
PV modules	3445	20	0.48

The operation and maintenance costs of the new renewable production units are €0.007/kWh [8]. The annual budget is €150,000 and the maximum portfolio investment for the life time of the devices is €5,500,000. Data on demand, wind speed, and PV factors used per time block are shown in Table 3, where wind and PV production levels are also displayed. The different levels are combined with each other in every time block. There are four levels of demand, wind, and PV production, hence, the total number combinations (scenarios) per time block is 64. Hence, there are 12 time blocks, 6 per season (winter (October–March) and summer (April–September)), accounting for 768 scenarios. Note that energy prices are not used in scenario generation and each price corresponds with a defined demand factor. These prices are shown in Table 3 and increase 1% each year with respect to the base year. All the levels in each time block have a probability of 1/4. Therefore, the weight of each of the scenarios within each time block is 1/64 [22]. Two blocks are used for piecewise linearization, where the cost of energy not supplied is €15,000/MWh. The tolerance ε of the Benders' algorithm is specified as €1.

Table 3. Investment data.

Blocks	Hours	Price [€/MWh]	Demand Factors	Wind Speed Factors	Wind Production Factors	Irradiation Factors	PV Production Factors
1	144	70.99	0.92	0.45	1.00	0.78	0.76
		62.89	0.90	0.29	0.76	0.20	0.20
		67.36	0.88	0.17	0.37	0.00	0.00
		57.74	0.87	0.05	0.00	0.00	0.00
2	1108	53.41	0.85	0.49	1.00	0.71	0.70
		48.54	0.82	0.29	0.76	0.17	0.17
		42.94	0.80	0.20	0.45	0.00	0.00
		44.85	0.78	0.08	0.08	0.00	0.00
3	960	45.59	0.77	0.47	1.00	0.76	0.74
		41.21	0.75	0.29	0.77	0.21	0.21
		42.09	0.72	0.19	0.42	0.00	0.00
		42.55	0.69	0.07	0.04	0.00	0.00
4	1029	41.65	0.66	0.48	1.00	0.73	0.71
		39.94	0.62	0.29	0.76	0.22	0.22
		37.11	0.59	0.19	0.44	0.00	0.00
		35.13	0.55	0.08	0.07	0.00	0.00
5	1027	32.37	0.52	0.49	1.00	0.72	0.70
		28.50	0.50	0.28	0.73	0.16	0.16
		23.91	0.49	0.19	0.42	0.00	0.00
		25.36	0.47	0.07	0.04	0.00	0.00
6	112	18.87	0.45	0.43	1.00	0.76	0.74
		16.10	0.44	0.24	0.62	0.18	0.18
		18.35	0.43	0.15	0.30	0.00	0.00
		3.19	0.40	0.05	0.00	0.00	0.00
7	46	57.38	0.97	0.47	1.00	0.76	0.74
		55.58	0.95	0.32	0.85	0.48	0.48
		54.79	0.94	0.19	0.45	0.08	0.08
		53.64	0.93	0.08	0.06	0.00	0.00
8	1083	53.19	0.89	0.44	1.00	0.69	0.67
		52.62	0.85	0.27	0.71	0.27	0.27
		49.51	0.82	0.18	0.41	0.01	0.01
		48.08	0.80	0.07	0.04	0.00	0.00

Table 3. *Cont.*

Blocks	Hours	Price [€/MWh]	Demand Factors	Wind Speed Factors	Wind Production Factors	Irradiation Factors	PV Production Factors
9	1084	44.25	0.78	0.45	1.00	0.70	0.69
		47.11	0.76	0.28	0.73	0.31	0.31
		44.67	0.73	0.19	0.44	0.02	0.02
		46.98	0.70	0.08	0.07	0.00	0.00
10	1028	47.89	0.67	0.46	1.00	0.71	0.70
		47.16	0.65	0.28	0.74	0.30	0.29
		43.81	0.62	0.19	0.43	0.01	0.01
		42.42	0.58	0.07	0.05	0.00	0.00
11	1025	41.53	0.55	0.47	1.00	0.69	0.67
		39.15	0.53	0.28	0.75	0.26	0.26
		38.14	0.51	0.20	0.46	0.00	0.00
		34.68	0.49	0.08	0.09	0.00	0.00
12	114	34.74	0.47	0.45	1.00	0.64	0.63
		36.49	0.47	0.28	0.73	0.20	0.20
		32.55	0.46	0.19	0.44	0.00	0.00
		33.54	0.44	0.08	0.07	0.00	0.00

6. Results Discussion

Two case studies have been simulated to test the model, where the constraints related to installed power and limits of investment are different. The results of Benders' algorithm for each case study case are compared with the MILP model given in Equations (1)–(56). These results are obtained using CPLEX 11 under GAMS [23] on an Intel Xeon E7-4820 computer with 4 processors at 2 GHz and 128 GB of RAM. Table 4 presents the numerical results of each simulation in each study case for 768 scenarios. The relative gap is set to 0.01 in all simulations for Benders' simulation.

Table 4. Total system costs (€).

Case	a		b	
	MILP	**Benders**	**MILP**	**Benders**
Total Costs	12,330,870	12,336,090	11,748,580	11,758,350
O&M system costs	11,178,621	11,134,070	9,308,298	9,314,627
Investment costs	1,152,249	1,202,020	2,440,282	2,443,723
CPU time (hours)	742	15	444	24.8

- Case a. Investment limits included: This case represents the most realistic scenario all constraints of the model are taken in account. The first year of the time horizon the renewable technology chosen for investment is photovoltaic (PV), with an installed capacity of 1047.5 kW (see Table 5). The expansion of the substation is carried out in the ninth year, with a new transformer. The new PV devices are installed at the end of the branches because it reduces the costs associated with energies losses. The location of the PV modules within the network is displayed in Table 6. The CPU time decreases by 97.9% when Benders' decomposition is the method chosen for the simulations.

Table 5. Power installed (kW) in Case a.

Case		a	
Year	**Substation**	**Wind**	**PV**
1	-	-	1047.5
9	1000	-	-
Total		**2047.5**	

Table 6. Nodal allocation of new power installed for Case a.

Year	Substation		Wind		Photovoltaic	
	Node	No. Units	Node	No. Units	Node	No. Units
					11	77
					24	85
1	-	-	-	-	25	85
					26	85
					27	85
					31	2
9	1	1	-	-	-	-

- Case b. Investment limits not included: in this case, investment constraints (Equations (16) and (17)) are not considered. This new constraint scenario allows the investment in, not only new transformers and PV modules, but also in wind units (see Table 7). Two expansions of the substation are made, in years 9 and 16, PV modules are installed in all candidate nodes, and six wind units, in total, are also installed in the last year. In this case, the new installed capacity is 4725 kW. The location of the PV modules within the network is displayed in Table 8. The CPU time decreases by 94.4% using Benders' algorithm.

Table 7. Power installed (kW) in Case b.

Case		b	
Year	Substation	Wind	PV
1	-	-	2125
9	1000	-	-
16	1000	-	-
20	-	600	-
Total		4725	

Table 8. Nodal allocation of new power installed for Case b.

Year	Substation		Wind		Photovoltaic	
	Node	No. Units	Node	No. Units	Node	No. Units
					11	85
					12	85
					24	85
					25	85
1	-	-	-	-	26	85
					27	85
					31	85
					32	85
					33	85
					34	85
9	1	1	-	-	-	-
16	1	1	-	-	-	-
			21	2		
20	-	-	22	2	-	-
			23	2		

The results are better in Case b than Case a because the problem is less constrained. The introduction of renewable energy in distribution systems reduces the total operation and maintenance (O&M) system costs (see Table 9). The reduction of O&M costs is due a reduction of losses costs and purchase energy by substation. These results highlight the advantages of investing in renewable generation in the long term.

Table 9. Total O&M system costs (€) for 768 scenarios.

Case	a	b	a	b
	MILP	**Benders**	**MILP**	**Benders**
Total O&M Costs	**11,178,621**	**11,134,070**	**9,308,298**	**9,314,627**
Losses costs	701,609	694,314	598,504	597,734
Not supplied energy cost	13,572	54,010	361	14,866
Purchase energy cost	10,195,660	10,104,480	8,138,853	8,130,402
DG O&M costs	267,780	281,266	570,580	571,625

Finally, Table 10 shows the CPU times required to solve the DGP problem using both the MILP model (1)–(56) and the Benders' algorithm for different number of scenarios. The CPU time required to solve the MILP model (1)–(56) increases drastically with the number of scenarios (Figures 3 and 4). However, its increment is approximately linear if Benders' algorithm is considered. This causes that, up to 64 scenarios, MILP model solves the problem faster than Benders' algorithm but, from 216 scenarios on, Benders' algorithm becomes a much more efficient way to solve it.

Table 10. CPU Time.

Problem	Case a					Case b				
Scenarios	1	8	64	216	768	1	8	64	216	768
MILP	4 s	1.9 m	0.7 h	13 h	742 h	1 s	30 s	3 m	11.1 h	444 h
Benders	36 s	5.3 m	1.7 h	4.9 h	15 h	11 s	90 s	72 m	5.1 h	24.8 h

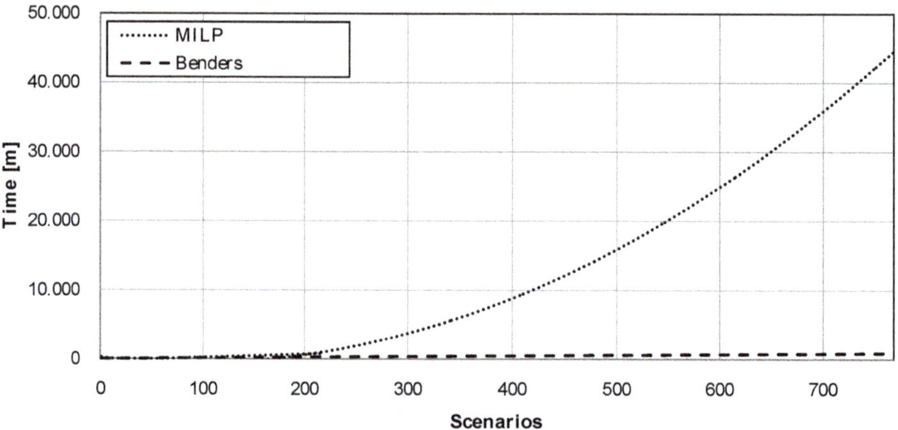

Figure 3. CPU time for Case a.

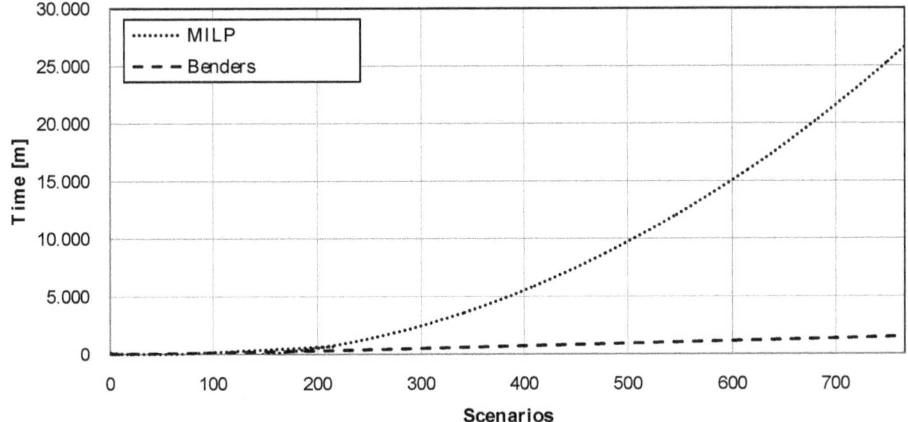

Figure 4. CPU time for Case b.

7. Conclusions

This paper has considered the DGP problem in a stochastic environment, where both PV and wind technologies as well as demand are subject to random changes. The use of Benders' decomposition to solve the two-stage stochastic investment problem has allowed us to further decompose the problem by scenarios and planning periods, making it a fully decomposable one. The model has been tested for a 34-bus example with excellent results.

In terms of computing time, the increase in the number of scenarios makes the differences between MILP and Benders' models evident. Up to 64 scenarios, the MILP model is much faster. Benders' model needs to perform iterative processes (loops) that increase the computing time. This trend changes as the number of scenarios increases and Benders' method becomes faster than MILP. In general, MILP's computing time behaves in a non-linear way, whereas Benders' model is more linear in terms of computing time. This proves the significant computational advantage of Benders' with respect to a conventional MILP model.

Note that applying Benders' decomposition may also allow extending the investment problem to address other relevant issues in distribution systems, such as switching or network reconfiguration. This would be non-viable using the original mixed-integer linear programming problem due to its high computational burden.

The work developed in this article can help investors to decide the kind and size of renewable technologies and the place where install the new devices in the distribution system. As improvements to the proposed problem, other kinds of producers (biomass, hydraulic) can be introduced, incorporate reliability in generation and electric vehicle. In addition, making a comparative with other uncertainty management methods such as K-means.

Author Contributions: Conceptualization, S.M.-B., J.I.M.-H., J.C. and L.B.; methodology, S.M.-B., J.I.M.-H., J.C. and L.B.; software, S.M.-B.; validation, J.I.M.-H., J.C. and L.B.; formal analysis, S.M.-B., J.I.M.-H., J.C. and L.B.; investigation, S.M.; resources, S.M.-B., J.I.M.-H., J.C. and L.B.; data curation, S.M.-B.; writing—original draft preparation, S.M.-B.; writing—review and editing, J.I.M.-H., J.C. and L.B.; visualization, S.M.-B.; supervision, J.I.M.-H., J.C. and L.B.; project administration, J.C.; funding acquisition, J.C.; J.I.M.-H., and L.B. All authors have read and agreed to the published version of the manuscript.

Funding: This research was funded by the Ministry of Science, Innovation, and Universities of Spain under Projects ENE2015-63879-R, RTI2018-096108-A-I00 and RTI2018-098703-B-I00 (MCIU/AEI/FEDER, UE), and the Junta de Comunidades de Castilla—La Mancha, under Project POII-2014-012-P.

Conflicts of Interest: The authors declare no conflict of interest.

References

1. Payasi, R.P.; Singh, A.K.; Singh, D. Review of distributed generation planning: Objectives, constraints and algorithms. *Int. J. Eng. Sci. Technol.* **2011**, *3*, 133–153. [CrossRef]
2. Viral, R.; Khatod, D.K. Optimal planning of distributed generation systems in distribution system: A review. *Renew. Sustain. Energy Rev.* **2012**, *16*, 5146–5165. [CrossRef]
3. Hadjsaid, N.; Canard, J.F.; Dumas, F. Dispersed generation impact on distribution networks. *IEEE Comput. Appl. Power* **1999**, *12*, 23–28. [CrossRef]
4. Barker, P.P.; De Mello, R.W. Determining the impact of distributed generation on power systems: Part I. Radial distribution systems. *IEEE Power Eng. Soc. Summer Meet.* **2000**, *3*, 1645–1656.
5. Dugan, R.C.; McDermott, T.E.; Ball, G.J. Planning for distributed generation. *IEEE Ind. Appl. Mag.* **2001**, *7*, 80–88. [CrossRef]
6. El-Khattam, W.; Hegazy, Y.G.; Salama, M.M.A. An integrated distributed generation optimization model for distribution system planning. *IEEE Trans. Power Syst.* **2005**, *20*, 1158–1165. [CrossRef]
7. Keane, A.; O'Malley, M. Optimal allocation of embedded generation on distribution networks. *IEEE Trans. Power Syst.* **2005**, *20*, 1640–1646. [CrossRef]
8. Zou, K.; Agalgaonkar, A.P.; Muttaqi, K.M.; Perera, S. Multi-objective optimization for distribution system planning with renewable energy resources. In Proceedings of the ENERGYCON 2010: IEEE International Energy Conference, Manama, Bahrain, 18–21 December 2010; pp. 670–675.
9. Kagan, N.; Adams, R.N. A Benders' decomposition approach to the multi-objective distribution planning problem. *Int. J. Electr. Power Energy Syst.* **1993**, *15*, 259–271. [CrossRef]
10. Zakariazadeh, A.; Jadid, S.; Siano, P. Economic-environmental energy and reserve scheduling of smart distribution systems: A multiobjective mathematical programming approach. *Energy Convers. Manag.* **2014**, *78*, 151–164. [CrossRef]
11. Ghasemi, A.; Mortazavi, S.S.; Mashhour, E. Integration of nodal hourly pricing in day-ahead SDC (smart distribution company) optimization framework to effectively activate demand response. *Energy* **2015**, *86*, 649–660. [CrossRef]
12. Hytowitz, R.B.; Hedman, K.W. Managing solar uncertainty in microgrid systems with stochastic unit commitment. *Electr. Power Syst. Res.* **2015**, *119*, 111–118. [CrossRef]
13. Bloom, J.A.; Caramanis, M.; Charny, L. Long-range generation planning using generalized Benders' decomposition: Implementation and experience. *Oper. Res.* **1984**, *32*, 290–313. [CrossRef]
14. Kim, H.; Sohn, H.S.; Bricker, D.L. Generation expansion planning using Benders' decomposition and generalized networks. *Int. J. Ind. Eng.* **2011**, *18*, 25–39.
15. Wang, J.; Wang, R.; Zeng, P.; You, S.; Li, Y.; Zhang, Y. Flexible transmission expansion planning for integrating wind power based on wind power distribution characteristics. *J. Electr. Eng. Technol.* **2015**, *10*, 709–718. [CrossRef]
16. Kazempour, S.J.; Conejo, A.J. Strategic generation investment under uncertainty via Benders' decomposition. *IEEE Trans. Power Syst.* **2012**, *27*, 424–432. [CrossRef]
17. Baringo, L.; Conejo, A.J. Wind power investment: A Benders' decomposition approach. *IEEE Trans. Power Syst.* **2012**, *27*, 433–441. [CrossRef]
18. Montoya-Bueno, S.; Muñoz, J.I.; Contreras, J. A stochastic investment model for renewable generation in distribution systems. *IEEE Trans. Sustain. Energy* **2015**, *6*, 1466–1474. [CrossRef]
19. Meneses de Quevedo, P.; Contreras, J.; Rider, M.J.; Allahdadian, J. Contingency assessment and network reconfiguration in distribution grids including wind power and energy storage. *IEEE Trans. Sustain. Energy* **2015**, *6*, 1524–1533. [CrossRef]
20. Conejo, A.J.; Castillo, E.; Mínguez, R.; García-Bertrand, R. *Decomposition Techniques in Mathematical Programming. Engineering and Science Applications*; Springer-Verlag: Heidelberg, Germany, 2006.
21. Kirmani, S.; Jamil, M.; Rizwan, M. Optimal placement of SPV based DG system for loss reduction in radial distribution network using heuristic search strategies. In Proceedings of the 2011 International Conference on Energy Automation, and Signal (ICEAS), Bhubaneswar, India, 28–30 December 2011; pp. 1–4.

22. Baringo, L.; Conejo, A.J. Correlated wind-power production and electric load scenarios for investment decisions. *Appl. Energy* **2013**, *101*, 475–482. [CrossRef]
23. Brooke, A.; Kendrick, D.; Meeraus, A.; Raman, R. *GAMS/CPLEX: A User's Guide*; GAMS: Washington, DC, USA, 2003.

Review

A Survey on Enhanced Smart Micro-Grid Management System with Modern Wireless Technology Contribution

Lilia Tightiz, Hyosik Yang * and Mohammad Jalil Piran

Department of Computer Engineering, Sejong University, 209, Neungdong-ro, Gwangjin-gu, Seoul 05006, Korea; liliatightiz@sju.ac.kr (L.T.); piran@sejong.ac.kr (M.J.P.)
* Correspondence: hsyang@sejong.edu; Tel.: +82-02-3408-3840

Received: 12 March 2020; Accepted: 23 April 2020; Published: 4 May 2020

Abstract: Micro-grid (MG) deployment has dramatically become more popular with the high penetration of renewable energy resources (RER). This trend brings with it the merits of independent power grid clean energy resource-based systems, and simultaneously the demerits of an unstable grid due to the intermittent nature of RER. Control and monitoring of MG through robust and ubiquitous communication system infrastructure is the solution to overcoming this intermittency. There has been an increasing focus in recent years on using wireless communication technologies as a prominent candidate in holistic proposal for the micro-grid management system (MGMS). The MGMS has been developed using the multi-agent system (MAS), multi-micro-grid (multi-MG), Internet of things (IoT) integration, and cloud concepts requiring new communication specifications, which can be satisfied by next-generation wireless technologies. There is, however, a lack of comprehensive corresponding investigation into management levels of MG interaction requirements and applied communication technologies, as well as a roadmap for wireless communication deployment, especially for the next generation. In this paper, we investigate communication technology applications in the MG and focus on their classification in a way that determines standard gaps when applying wireless for MG control levels. We also explore and categorize the literature that has applied wireless technologies to MG. Moreover, since MGMS has been revolutionized by attaching new technologies and applications to make it an active element of the power system, we address future trends for MGMS and offer a roadmap for applying new enhancements in wireless technologies, especially the fifth generation (5G) of wireless networks with its exclusive characteristics, to implement this novel approach.

Keywords: micro-grid; energy management system; IEC 61850; 5G; LoRa; multi-agent system

---

1. Introduction

Due to problematic issues with fossil fuels such as the limited resources, increasing greenhouse gases, and air pollution, new resources of energy, including solar, wind, tidal, etc. have been introduced, known as renewable energy resources (RER), and have overcome environmental issues with fossil fuels [1]. One of the most prominent advantages of RER is the end of dependency on conventional power plants and a centralized electricity network, which makes them sustainable candidates for distributed electricity generation, particularly in remote places. This specification, along with the development of communication technologies that facilitate information exchange in remote places to cover control, protection, and administration requirements of this distributed grid, have led to increasing RER use in the electricity grid, especially in the form of micro grids (MG) [2]. Based on the Institute of Electrical and Electronics Engineers (IEEE) standard 2030.7–2017 [3], MG is defined as a "group of interconnected loads and distributed energy resources with clearly defined electrical boundaries that act as a single controllable entity concerning the grid and can connect and disconnect from the grid to enable it to operate in both grid-connected or island modes". MG penetration is increasing, as it will decrease the cost of electricity transmission infrastructure by using RER. Corresponding to the definition of MG, there are two performance modes: islanded mode, and grid-connected mode. In islanded mode, MG should be able to work as an independent grid and supply consumers alone. On the other hand, in grid-connected mode, the connection of MG to a utility grid defeats the risk of unavailability owing to natural resource features, and offers other benefits related to participation in the electricity market as a prosumer, specifically selling over-produced electricity or buying it in the case of resource unavailability or system failure. All these advantages would not be viable without a robust communication system.

There has been a plethora of research about the optimal control, protection and management of MG [4–7], which have revealed the role of automation and smartization to achieve above-mentioned MG profit to be undeniable; however, communication structure and specification of the system are the key factors of all these scenarios [8–10]. A large number of possible communication protocols and configurations have been applied in MGMS depending on the control and protection scenarios of the system, geographical location of MG, initial investment and maintenance budget, the importance of loads, number of distributed energy resources (DER), traffic rates of the communication system, and so on. Many authors have attempted to find the best communication structure for international standards, such as the International Electrotechnical Commission (IEC) 61850, IEEE 1547, and IEC 61400–25 [11–14]. IEC 61850 is a standard of the communication protocol of automation in the power industry. This standard considers communication within power system substations and uses an open system interaction (OSI) model as a communication stack for data exchange in the local area network (LAN) in its first version. It has been developed for the intelligence of the entire breadth of the power system and interacts with MG by adding IEC 61850–7–420, IEC 61850–90–7, IEC 61850–8–2, and IEC 61850–9–12. These new parts specify information models for exchange information with DER along with the definition of communication stacks for wide-area network (WANs). IEC 1547, which is a standard of interconnection between DER and electric power systems, deals with safety, protection, power quality, and information exchange requirements of DER, as well as alternative configurations for MGMS. IEC 61400–25, which is a standard of communications for the monitoring and control of wind power plants, provides uniform model information for exchanging data. These standards facilitate control, protection, and management modeling of MG. In the case of communication technology deployment, studies have categorized MG interactions into two groups: wired and wireless. While RS-232, RS-485, Power Line Carrier (PLC), optical fiber, and ethernet are commonly used wired technologies in MG communications, ZigBee, Bluetooth, Wi-Fi, WiMAX, Global System for Mobile (GSM), 5G/4G/3G/HSPA, LoRa, and satellite communication are wireless ones [15–20]. Due to wireless technology advantages in comparison with wired technology, such as cost effectiveness, convenient installation, portability, low risk of ground potential issues, and scalability, an investigation into employing this technology in MGMS should be undertaken.

1.1. Related Works

Protection, control, management, and communication requirements, as well as the intermittent characteristic of RER as the main element of MG, has made this research area a trend in the power industry. This trend is divided into two different domains: electrical and communication. The performance of the predecessor is dependent on the successor. As shown in Table 1, several researchers have attempted to recognize communication architecture and technologies adapted to MG requirements as an interactive smart element of the public electricity grid. Safdar et al. [21] divided communication in MG into three levels: home area network (HAN), field area network (FAN), and WAN. The authors proposed appropriate communication technologies for each level. Based on the requirements of physical connections, they made wireless technology attractive for communication in MG. The authors in this paper mentioned reliability, quality of service (QoS), security, complexity, standardization, and efficiency as challenges in the communication infrastructure of MG while disregarding enough intention to standard protocols, and their relevant constraints, which have had a significant effect in communication technology employment. MG testbed structures and applied communication technologies were reviewed comprehensively in [22]; however, this review focused on characteristics of implemented structure communication technologies, and future trend guidelines were missing. Bani-Ahmad [23] then clarified the data flow and communication protocols and standards of MG, such as the Internet Protocols Suite, Modbus, DNP3, and IEC 61850, and focused on MG physical communication links in two parts: wired physical links and wireless physical links. Although this survey satisfied the protocol critical characteristics of communication in MG in comparison with typical IT systems, there was no classification of the application of highlighted technologies in MG communication structure. The role of communication in MG, wired and wireless technologies, and their limitation in MG applications were analyzed in [16]. Mavrokefalidis et al. [24] provided a brief overview of the communication infrastructure and challenges of applying wireless communication in MG. Marzal et al. [20] investigated the feasible network topology for smart MG along with MG evolution. The authors then noted smart MG issues, including bandwidth, latency, reliability, and cybersecurity, and studied communication protocols and technologies through the literature. The categorized communication technologies based on application and reviewed their characteristics and issues, as discussed in [25] and studied in [26] after the comparison of specifications of wired and wireless communications, which concentrated on satisfying communication requirements and the architecture of smart inverters in MG.

Table 1. Present and related works—a comparison.

Reference	Year	MG Communication Specification Requirements			MG Communication Technologies		Future Trends	Roadmap
		Standard	Control	Application	Wired	Wireless		
S.Safdar et al. [21]	2013	✗	✗	✓	✗	✓	✗	✗
L.Mariam et al. [22]	2013	✓	✗	✗	✓	✓	✓	✗
Bani-Ahmed et al. [23]	2014	✓	✗	✗	✓	✓	✓	✗
P.D.Chavan et al. [16]	2016	✓	✓	✓	✓	✓	✗	✗
C.Mavrokefalidis et al. [24]	2017	✗	✗	✗	✓	✓	✓	✗
S.Marzal et al. [20]	2018	✓	✓	✓	✓	✓	✓	✗
T.T.Mai et al. [25]	2018	✓	✓	✓	✓	✓	✗	✗
B.Arbab-Zavar et al. [26]	2019	✗	✗	✓	✓	✓	✓	✗
Present Paper	2020	✓	✓	✓	✓	✓	✓	✓

All these attempts investigated communications requisite for and technologies applied in MG and their specifications and applications, although some of the following aspects, which affect communication requirements in MG, have been neglected. Applying energy storage systems (ESS), RER, and responsive loads (RL) along with the introduction of new concepts in a smart grid (SG) and MGMS architecture such as MAS, multi-MG, and IoT integration, make MG an integrated part of the main grid. In this scenario, MG cooperates in the electricity market as a prosumer and supports the main grid by providing ancillary services (AS) such as demand response (DR), power system frequency, and voltage stability, as well as black-start aid and electricity market contribution by using technologies such as vehicle-to-grid (V2G), smart inverters, and so on. Moreover, communication protocol standards, which are applied in MG communication, need revising to attain appropriate a communication infrastructure for these new technologies. To produce these aspects, we focus on MG as SG tiles, and cut short the path to the development of SG penetration [27]. We will provide a supervisory hierarchical structure for MG corresponding to the SG conceptual model, as well as clarifying the situation of this structure in the distribution management system (DMS). Another idea behind this hypothesis is using MG to help power system stability by splitting the power grid into islands [28]. Table 1 shows how this paper, in comparison with related works, involves all examined MG control structures and wireless technology applications, in order to detect a roadmap for exploiting these technologies.

1.2. Contribution

To address the above-mentioned issues, the main contributions of this paper are summarized as follows:

- Investigation of MG communication requirements based on MGMS and DMS requirements.
- Study of relevant communication standards from the perspective of smart MG and standard revision requirements in a way facilitating wireless deployment in MG communication.
- Study of the deployment of wireless communication applications in MG through the classification of literature based on applications and methods.
- Determination of wireless technology applications in MG concerning communication requirements and standards of each application.
- Investigation using new wireless technologies such as 5G to attain a roadmap for wireless application in new MGMS trends such as MAS and multi-MG.

The remainder of this paper is organized as follows. Section 2 is about the concept of MG focusing on electrical and communication structures and specifications. Section 3 introduces communication standard protocols and their revision requirements to facilitate wireless deployment in MGMS. Communication technologies and their applications in each supervisory level of the MGMS are discussed in Section 4. To provide a roadmap for using new technologies in MG focusing on wireless technology and more specified enhanced ones such as 5G in Section 5, we classified the literature based on its objectives, MG application, and method qualification when using wireless technologies in MG environments. After opening new horizons for our paper research area in Section 6, our attempt is finalized in Section 7.

2. MG Concept and Requirements

2.1. MG Structure

MG, according to the definition proposed in the introduction to the paper, includes DER, which can be RER as conventional resources, ESS as an infrastructure of energy market contribution, or AS in the absence of DER, loads, and finally a breaker at the point where MG connects to the main grid. This collection of elements has two modes of operation, specifically, grid-connected and islanded mode. In grid-connected mode, they receive services or provide services for the main grid, whereas in islanded mode, they should be able to manage all control, stability, and protection requirements. Achieving these characteristics requires a smart structure for MG, and its elements should be developed with these requirements in mind. It is noted that the penetration and integration of smart MGs should be in the form of tiles facilitating the implementation of SG. Figure 1 is a visualization of this idea, where MG structure meets the conceptual model of SG. SG has seven sectors [28]: Customer, Service Provider, Transmission, Distribution, Bulk Generation, and Market. Adapting the MG structure to SG structure, while the role of generation, transmission, distribution, operator, and consumer is clear, as can be seen in Figure 1, the maintainer acts as a service provider and the aggregator is in the market domain. The performance of each entity is clarified here.

- **MG Aggregator:** This participant is in charge of gathering information about energy marketing participants for MG.
- **MG Operator:** This party monitors MG and controls its performance through local and remote services.
- **MG Maintainer:** This agent is responsible for the accurate performance of MG by providing maintenance services in case of failure according to the received reports.
- **MG Controller:** Aggregator, Maintainer, and operator interact through this part with each other and DER, facilitating the Energy Management System (EMS) control level.

- **Local Utility**: Local utility is the MG utility neighbor who connects to MG through the Point of Common Coupling (PCC) and interacts with MG to coordinate the provision of AS in grid-connected mode.

As well as embedding MG in SG domain, Figure 1 conveys a hierarchical control strategy of smart MG by introducing three supervisory levels, namely advanced metering interface (AMI), EMS, and DMS.

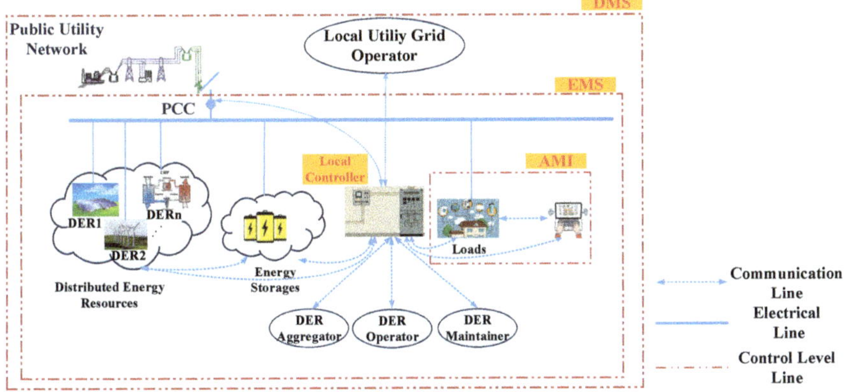

Figure 1. MG architecture.

- **AMI**: The first control level is related to load control. In this paper, this control level is considered in the scope of MG. Energy consumption of smart appliances is monitored through smart meters and the AMI control level is responsible for sending this information to the higher level, which is EMS. AMI makes it possible to cooperate with RL in load shedding, peak shaving, and other services.
- **EMS**: The second control level of MG is EMS, which is responsible for control demand and supply in MG based on information receives from DER, ESS, and loads. EMS has a key role in MGMS and provides a schedule for each MG element operation through the optimization algorithm in both grid-connected and islanded modes of MG. EMS decides MG contribution in AS provision for the utility grid in a way that determines the contribution portion of each source.
- **DMS**: The third control level is responsible for the control, monitoring, and reliability of the distribution network. This supervisory control level reveals when MG will be able to work in grid-connected and supply loads from the main grid or provide AS for utility and, in case of any failure in MG, it will disconnect from the main grid. There could be several MGs in DMS territory as shown in Figure 2.

The hierarchical control level of MG can be implemented through a centralized or decentralized approach. In the centralized approach, each element receives setpoints from a central controller and follows a global objective, while in decentralized approach decision making in each level control is distributed. In distributed fashion, each element has been considered to be an agent which independently decides to participate in SG or MGMS. This offers advantages such as plug-and-play patterns, no need for a dedicated communication infrastructure, fast response to system failure, and power grid consisting of independent entities [29,30]. Moreover, due to the significant advantages of MGMS characteristics when deploying MAS, trends toward the implementation of SG in the form of multi-MG have been augmented recently. It is estimated that future SG would be multi-MG that works in the MAS environment [31,32]. Figure 1 proposes the control level of individual MG without centralized and decentralized approach. The local controller in this figure can be considered to be located in EMS, or distributed in each element.

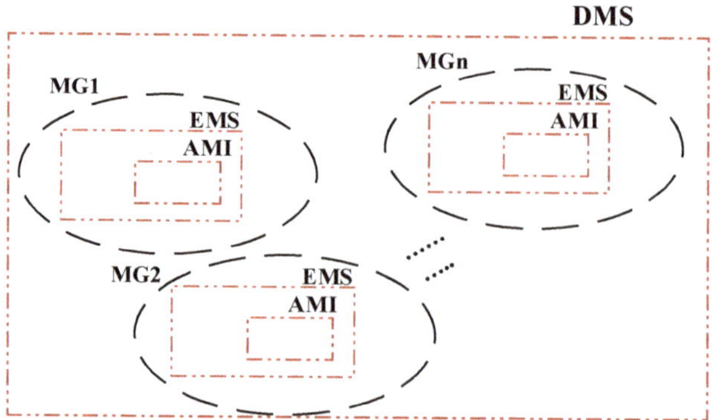

Figure 2. DMS supervisory level domain.

2.2. MG Communication Requirements

To detect efficient and appropriate communication technology for MG, we need to ascertain the specifications and constraints of information exchange in each supervisory level. Figure 3 shows a comparison of AMI, EMS, and DMS characteristics based on data rates, reliability, coverage, delay, complexity, security, and benefits per cost. This figure ensures each control level of MG has its communication requirements. While all control levels require a highly reliable and secure communication, their data rates and delay characteristics increase according to their coverage characteristic. The other important specification for communication infrastructure is backup power, which ensures the performance of communication during any outage and is proportional to the critical mission of each level control from zero for AMI, as interacting with meters increases to one hour for EMS, and 72 hours for DMS. In the SG environment, there is a hierarchical communication level including HAN, FAN, and WAN. This classification can be followed by MG. While HAN contains loads such as smart appliances and electric vehicles (EV), which can be automated and programmable through home energy management systems (HEMS) to provide AS for utility grid such as DR, FAN coordinates among DER, ESS, operators, and energy marketing entities, hence needing higher bandwidth than HAN. At the highest level, WAN embraces information exchange between EMS and DMS when MG works in grid-connected mode. Communication specifications of applications in MG are summarized in Table 2 [33].

Table 2. Communication specification application in MG.

Communication Level	Application	Bandwidth	Latency
	HEMS	9.6–56 kbps	200 ms–2 s
HAN	EV Charging	9.6–56 kbps	2 s–5 min
	V2G	9.6–56 kbps	2 s–5 min
FAN	AMI	node:10–100 kbps backhaul: 500 kbps	2–15 s
	DER and ESS	9.6–56 kbps	20 ms–15 s
WAN	DR	14–100 kbps	500 ms–several min
	DMS	9.6–100 kbps	100 ms–2 s

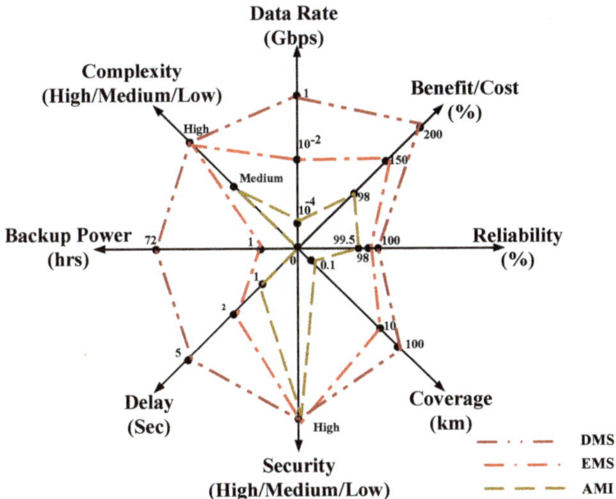

Figure 3. Comparison of MG control levels characteristics.

3. MG Communication Protocols and Standards

There is a plethora of standards that embody all aspects of MG electrical and communication requirements. Table 3 depicts all communication-relevant standards of MG, which have been clarified in the following.

Table 3. MG management- and communication-relevant standards.

	Standard Title	Reference	Title
IEC	61850–7–420	[34]	Basic communication structure–Distributed energy resources logical nodes
	61850–8–2	[35]	Specific Communication Service Mapping (SCSM)—Mapping to eXtensible Messaging Presence Protocol (XMPP)
	61850–90–7	[36]	Object models for power converters in distributed energy resources (DER) systems
	61850–90–12	[37]	Wide-area network engineering guidelines
	61400–25–2	[38]	Communications for monitoring and control of wind power plants–Information models
IEEE	1547	[39]	IEEE Standard for Interconnection and Interoperability of Distributed Energy Resources with Associated Electric Power Systems Interfaces
	2030	[40]	Smart Grid Interoperability of Energy Technology and Information Technology Operation with the Electric Power System (EPS), End–Use Applications, and Loads

3.1. IEC Standards

3.1.1. IEC 61850

IEC 61850 is an international standard of communication networks and systems for power utility automation. This standard in the 2003 edition focused on the automation of power system substations and was divided into three levels: Process, Bay, and Station. In the process level, measurement equipment such as current transformers (CT), potential transformers (PT), or phasor measurement units (PMU) are located and send measurement data to the bay level, which is the place of intelligent electronic devices (IED) issuing control and protection commands based on the received information. The process level communicates with one or more bays, and even the whole substation, as well as exchanging information with the engineering workplace for remote control commands. All of this information exchange in a form of three types of the message, including manufacturing message specification (MMS) for non-critical information and in the format of request-response, generic object-oriented substation event (GOOSE) for critical information with high priorities such as trip command and sampled value (SV) for measurement units with high priority. GOOSE and SV are of multicast format. Each of these messages has its time limitation and is mapped according to the communication stack proposed in part 8–1 of the standard, which uses an open system interconnection (OSI) model and Ethernet as a physical layer in the LAN environment. In MG, an IED is located in each DER for the control and protection of objects. The nature of MG is that it is positioned in remote places desiring interconnection with the main grid through WAN. An extension of IEC 61850 in successive editions by information model supporting MG in part 7–420, 90—7, and providing a communication stack for interconnection in WAN based on applying eXtensible Messaging and Presence Protocol (XMPP) for mapping information in part 8–2 makes this standard appropriate for use in the MG communication environment. However, it is noted that other IoT protocols are suggested in other literature, and laboratory experiences perform better than XMPP in comparison with QoS, implementation infrastructure requirements, and future development specifications [41,42].

3.1.2. IEC 61400–25–2

This standard presents a uniform platform of information exchange and a data model for all participants in the wind power station, and serves as a communication stack for mapping this information. This standard deploys common data classes of IEC 61850 to be corporate in SG. Its data object has been extended in the recent version [43].

3.2. IEEE Standards

3.2.1. IEEE 1547

IEC 1547 is a series of standards facilitating operation, control, monitoring, and integration of MG for utility networking for the provision of AS. The main drawback of this standard is the detail of establishing communication of MG with the supervisory level of the power system, i.e., TSO/DSO is unclear [43,44].

3.2.2. IEEE 2030

This standard is a series of standards clarifying the SG Interoperability Reference Model (SGIRM). In part seven it proposes specification of MG controllers and represents MG control system functional framework [43,44].

4. MG Communication Technologies

In this section, to conform with the role of the hierarchical level of communication in MG, relevant technologies will be introduced. Technologies are partitioned into two groups: wired and wireless technologies.

4.1. Wired Technologies

Wired technologies, which can be used in MG, consist of PLC and ethernet at HAN and FAN level, Coaxial Cable and DSL in FAN level, and Fiber Optic in WAN level. Implementation of wired technologies costs more than wireless, especially in MG, since it is usually located in remote places, although they are not battery-dependent like wireless and are less affected by interference. However, the huge amount of interaction between sensors, actuators, meters, and controllers to provide smart MG makes it an inevitable technology. As our focus in this paper is wireless technologies, we just consider PLC, which is a legacy technology widely used for information communication in the power system for many years with low-cost implementation, as it uses transmission lines to interchange data by ultra-high frequency (UHF). This specification will be affected by the noise of other connected devices to the network. Ultra-Narrow Band PLC (UNBPLC), Narrow Band PLC (NBPLC), and Broad Band PLC (BBPLC), are three different categories of PLC deploying in MG. Wired technologies, characteristics, applications, and obstacles are represented in Table 4 [26,45].

4.2. Wireless Technologies

Trends to use wireless technologies in MG have been rising to reduce the complexity and cost of the system. Table 5 represents wireless technologies specification, namely data rates, coverage, application in MGMS, and merits and demerits of each technology. There are several wireless technologies, including Wireless Personal Area Network (WPAN), Wireless LAN (WLAN), cellular networks, Low Power Wide-Area Network (LPWAN), and Satellite Network within different standards applied in MG, of which noticeable ones will be discussed, following [21,26,45–49].

4.2.1. ZigBee

ZigBee is one of the widely used wireless technologies in HAN, especially in cases of the smart home because of low cost, energy consumption, and complexity working in the unlicensed band of 2.4 GHz. It is also one of the integrated communication technologies of metering equipment. ZigBee can provide tree, star, and mesh network topologies by Direct Sequence Spread Spectrum (DSSS) modulation. However, it suffers from common weak points of Industrial, Scientific, and Medical (ISM) license-free spectrum, which is at the risk of interference, although there are some approaches, such as the channel discovery algorithm based on Carrier Sense Multiple Access/Collision Avoidance (CSMA/CA) interference, that have been deployed to deal with this issue. It has been noted that in monitoring operation use cases, Zigbee is a prominent one.

4.2.2. WiFi

WiFi is a wireless technology used for WLAN implementation within the IEEE 802.11 standard. It provides a frequency spectrum with a data rate of 2-600 Mbps and covers up to 100 meters distance. High penetration of WiFi as the Internet infrastructure makes it a good candidate for HAN and FAN applications.

4.2.3. WiMAX

This technology is one of the communication standards of the IEEE 802.16 series operating in licensed spectrum 2.5 or 3.5 GHz and unlicensed spectrum of 5.8 GHz. Since there is along-distance coverage of licensed spectrum, it is appropriate for communication in HAN and FAN level in MG, especially in advanced AMI such as AMI for the supervision of measurement equipment in HAN. It can facilitate responsive load contribution in the provision of AS such as peak shaving or load shedding. It is reported that applying WiMAX in remote places requiring WiMAX tower installation, which will result in imposing an excess cost to the system.

Table 4. Wired technologies specification and application in MG communication structure.

Technologies	Standard/Protocols	Data Rate	Coverage	Application in MG Control	Advantages	Disadvantages
Coaxial Cable	DOCSIS	172 Mbps	28 km	Smart appliances, AMI	✓Low cost ✓Easy installation	✓Low scalability and bandwidth ✓Susceptible to noise
Ethernet	802.3X	10 Mbps–10 Gbps	100 m	Smart appliances, AMI	✓Well performance of capacity, reliability, availability, and security	✓Complexity of installation ✓Not perfect in real-time communication
DSL	HDSL	2 Mbps	3.6 km	AMI	✓Cost effective because of existing infrastructure	✓Possibility of degradation in data quality
	ADSL	1–8 Mbps	5 km		✓Proven technology in residential services	✓Out of power system management and supervisory
	VDSL	15–100 Mbps	1.5 km			
Fiber Optic	PON	155 Mbps–2.5 Gbps	60 km	AMI, DMS	✓Not affected by noise and electromagnetic interferences	✓High Cost
	AON	100 Mbps	10 km		✓Well performance of Capacity, Reliability, Availability, Security, and Latency	✓Low Scalability
	BPON	155–622 Mbps	20–60 km			
	GPON	155 Mbps–2.448 Gpbs	20 km			
	EPON	1 Gbps	20 km			
	SONET/SDH	10 Gbps	100 km			
	WDM	40 Gbps	100 km			
PLC	UNBPLC	100 bpc	150 km	Smart appliances, AMI, EMS, DMS	✓Convenience and cost effective because of no need separatedinfrastructure from power grid	✓Subjected by interference of power grid noise or weather conditions
	NBPLC	10–500 kbps	150 km			
	BBPLC	10–200 Mbps	1.5 km			

Table 5. Wireless technologies specification and application in MG communication structure.

Technologies	Standard/Protocols	Data Rate	Coverage	Application in MGMS	Advantages	Disadvantages
WPAN	Z-Wave	40 Kbps	30 m	Smart Appliances HEMS	✓No interference ✓Free bandwidth ✓Mesh Connectivity	✓Low data rate ✓high power consumption
	Bluetooth	1–2 Mbps	15–30 m	Smart Appliances HEMS	✓Free bandwidth ✓Low power consumption ✓High Data Rate	✓Very short range ✓Subjected by noise ✓Unsafe
	ZigBee	250 Kbps	100 m	HEMS EVs	✓Low cost ✓Low Power Consumption	✓Low data rate ✓Short Range ✓Interference
	WirelessHART	115 Kbps	200 m	HEMS Smart meters	✓Scalable ✓Backward compatible	✓Low data rate ✓Short range ✓Interference
	ZigBee Pro (inter-WPAN)	250 Kbps	1.6 km	V2G, AMI	✓Mesh Connectivity	✓Low data rate ✓Interference
WLAN	WiFi(802.11X)	2–600 Mbps	100 m	Smart Appliances V2G AMI	✓Low cost ✓High Data rate ✓Wide adoption	✓short range ✓Interference ✓Low Security
Cellular Network Communication	WiMAX(802.16)	75 Mbps	50 km	DMS EMS AMI DR	✓High Data rate ✓QoS Provisioning ✓Scalability ✓Low Latency	✓Not widespread use ✓Dedicated Infrastructure ✓Limited Access to licensed spectrum
	2G (GSM)	14.4 Kbps	1–10 km		✓Existing infrastructure and service models	✓Oriented for human broadband applications
	2.5G (GPRS)	144 Kbps	1–10 km		✓Ubiquitous coverage	
	3G	2 Mbps	1–10 km		✓Low latency	✓Monthly recurring charges
	3.5G	14 Mbps	0–5 km		✓High data rate	✓Not currently support for mission-critical applications
	4G	1 Gbps	50 km		✓QoS Provisioning	
	5G SIFOX	>1 Gbps 100 bps	50 km Urban Area: 3–10 km Rural Area: 30–50 km	AMI DMS	✓Low power consumption	✓Low data range
LPWAN	LoRa	LoRa modulation: 0.3–37.5 Kbps LoRaWAN: 50 Kbps	Urban Area: 2–5 km Rural Area: 10–15 km		✓better data range (LoRaWAN)	
Satellite Network	LEO	Iridium: 2.4–28 Kbps	100–6000 km	DMS, AMI	✓Wide-area coverage	✓High cost
	MEO	Inmarsat-B: 9.6–128 Kbps			✓High reliability	✓High Latency
	GEO	BGAN: 1 Mbps				

4.2.4. 5G

Cellular networks such as GSM, GPRS, 3G, 3.5G, 4G, and 5G have high data rate communication and bandwidth. This characteristic facilitates the transfer of a huge amount of data. Therefore, these technologies can be used in applications such as interchange among smart meters, MG control center, and supervisory level of the main grid, i.e., DMS in WAN environments. High-cost licensed spectrum and the uncertainty of stable connectivity in severe weather conditions are drawbacks of this technology, although cellular technologies could defeat interference and security issues of free bands by applying licensed bands. Cellular communication, which was initially introduced in the 1980s based on analogue signal communication, has been developed to IP-based communication such as 4G and 5G for increasing bandwidth and sustaining real-time communication. 4G, which is called Long-Term Evolution-Advanced (LTE-A), increased the data rate of its predecessor generation, i.e., 3G from the range of 2–14 Mbps to the range of 100 Mbps-1 Gbps as well as latency and improvement of infrastructure energy consumption. Meanwhile, noticeable advancement in this technology is happening by the introduction of 5G to facilitate IoT application penetration in MGMS. The high performance of 5G has three characteristics, including Millimeter-Wave (mm-Wave), Multiple-Input Multiple-Output (MIMO), and ultra-dense cellular network. This technology brings benefits such as lower latency, higher bandwidth, a larger number of participants and nodes in MGMS, higher security, and so on. Enhanced Mobile Broadband (eMBB), Ultra-reliable and Low-latency Communications (uRLLC), and Massive Machine Type Communications (mMTC) are services offered by 5G as The International Telecommunication Union (ITU) determined. While an eMBB service is a response to individual aspects of the world digitalization communication requirements, uRLLC and mMTC are related to higher scales of this trend, including industry and society aspects, respectively. Hence eMBB provides higher bandwidths for applications such as 3D and High-Definition (HD) Video and Virtual Reality (VR). This ultra-bandwidth can reach to 10 Gbps. On the other hand, uRLLC offers low latency limited to 1 ms for time and reliability sensitive applications such as driverless cars and industrial automation. Finally, mMTC represents the number of connected devices up to 1 million/km^2, facilitates the implementation of smart homes and smart cities [50].

4.2.5. LPWAN

This technology was introduced to provide equilibrium between energy consumption and data rate in WAN. Cellular networks have been widely used to transmit high data rates in WAN but with high energy consumption. LPWAN could defeat this weak point of a cellular network by introducing SIGFOX and LoRa. These technologies have been applied to a star topology, which will result in the simplicity of the system and less power consumption. LoRa technology is divided into two subsets, including LoRa modulation and LoRaWAN. LoRa is a physical layer protocol using ISM bandwidth modulation. Despite Frequency Shifting Key (FSK) used by legacy wireless protocols, LoRa applies Chirp Spread Spectrum (CSS) brought by a higher data rate for LoRa. While LoRa modulation is a physical layer protocol, LoRaWAN contains the MAC layer as well as the physical layer and improves data rate concerning power consumption [51–54].

4.2.6. Satellite Technology

Satellite technology is a solution for MG communication in remote places where cellular or other wireless technologies are not accessible. Furthermore, they can be applied as a redundant path for creating backup communication. There are three different satellite technologies, including Low Earth Orbits (LEO), Medium Earth Orbit (MEO), and Geostationary Earth Orbits (GEO). Table 5 shows the characteristics of satellite technologies. The major disadvantages of satellite technologies are high expense and high latency. However, LEO and MEO areas are being developed to improve latency and bandwidth. More service providers represent GEO, including Inmarsat, BGAN, Swift, and MPDS while just NEW ICO offers MEO. Iridium and Globalstar are operators of LEO [55,56].

5. Wireless Technologies Roadmap and Future Trends for MG

MG is one of the key elements of the developed power grid system, which is being widely used because of the aforementioned merits. Regarding wireless specification correspondence of MG remote and independent specification, in the previous section, features of wireless technologies and applications in each part of the hierarchical supervisory of MG are discussed. Table 6 depicts how the literature considers applying wireless in MG. As can be seen, the literature proves wireless communication benefits in MG according to different goals and examination methods. We classified them into control, standard, and technology-based categories. This table reveals a new approach to the role of MG in the power system, as an active element is under trial and error test by using new enhancements of wireless technologies [52,57]. Concerning countless merits of wireless technologies application in MG, it is necessary to determine a roadmap clarifying the future path of this enhancement. To provide a roadmap for applying wireless technologies in MGMS, we should consider the following aspects.

5.1. MGMS Structure Development

New wireless technologies should consider communication specification requirements of MG application such as bandwidth, latency, and security. To provide a mechanism for applying wireless in MG, we require a definition of unified information exchange and standard. The unified model should be flexible to apply in the different control structures of MG depending on the extent, application, and geographical location of MG. As discussed before, MG can be centralized or decentralized. One of the promising strategies in distributed control of MG is MAS, which facilitates control of MG and assists the SG concept approach in reality by accommodating self-healing characteristics for the power grid [7,58–61]. In the MAS environment, every element of MG can interact and cooperate with other elements intelligently to achieve power system global objectives, as can be seen in Figure 4. In this case, although there is no "one size fits all" approach, robust technologies such as 5G can be deployed in two control architectures. Central control of MG can derive active power and reactive power of load information and control generation portion of each power resource in MG by applying 5G communication and cloud computing at the edge. It is also possible for 5G employment in decentralized MG control based on droop control power, which needs real-time reference control signals, large distance coverage, and security [62]. As well as the independent grid characteristic of MG, there are new aspects in the smartization of MG, making it an integrated part of the public grid, which is the main hypothesis of this paper. MG as an AS of the main grid is one of the aspects, which will be an incentive for penetration extension of MG as it provides profit for owners. Such aspects require the contribution of technologies in MG such as V2G, smart inverter, etc. Wireless technology is a good candidate as a communication infrastructure of these technologies [63]. AS provision of the main grid, including frequency regulation, voltage control, and black-start aid by MG needs its communication characteristics, especially time constraints, which can be met by 5G mm-Wave characteristics offering fast, reliable, and robust interacting [64–67].

5.2. MGMS Communication Standard Development

The development of relevant standards introduced in Section 3 is a prerequisite to meet wireless technologies in MGMS. For instance, WirelessHART in [68] was noticed facilitating IEC 61850 in the distributed generation sector. A communication stack of IEC 61850, which is the OSI model, is changed and information model and exchange mapped to the wirelessHART protocol. However, it could not meet time limitation requirements of the standard, and the authors referred to solving this weak point to future work. This practice shows that since the most serving MGMS communication standard is IEC 61850, the communication stack of this standard should consider technology developments of wireless communication concerning their coordination with new trends in MGMS.

Table 6. Classification of wireless-based MG studies.

Classification	Ref	Wireless Technology	Standard/Protocol Deployment	Qualification	Tools	Objectives
Standard-based	[69]	WLAN	IEC 61850	✓GOOSE messages performance, time delay and throughput requirements, as well as the distance between DAS and DER	OPNET	✓Communication systems between IEC 61850-based distribution substation and DER
	[68]	WirelessHART	IEC 61850	-	OPNET	✓Cabling and installation cost reduction,Portability
	[70]	WiFi IEEE 802.11.n,Z-wave	IEC 61850	✓Latency, Availability, Time synchronization accuracy, Reliability	Real test	✓Automation of MG using heterogeneous communication
	[71]	WiFi IEEE 802.11.g	IEC 61850	✓Effect of communication distance, Delay in burst background Traffic	OPNET	✓Smart MG based on IEC 61850
	[72]	WiFi IEEE 802.11.g,	IEC 61850	✓Average delay	Riverbed modeler	✓Smart home and smart meters based on IEC 61850
		WiFi IEEE 802.11.n, WiMax 802.16		✓Packet loss per second		
Control-based	[73]	Cellular Network	-	✓Data rates, ranges, and capacity of the cognitive radio	-	✓Data service energy center by wireless access to AMI
	[74]	WiMAX	-	✓Bit error rate	MATLAB	✓Real-time protection,Reliability
	[75]	wireless community mesh network	-	✓Impact of the time-varying wireless communication delay on the performance of distributed power inverters	MATLAB	✓Load Sharing
	[76]	-	-	✓Maintains the magnitude of the voltage and frequency within the acceptable limits (EN 50160)	-	✓Wireless EMS
	[77]	IEEE 802.11 MAC standard	-	✓MAS-based decentralized MG control	NS-2	✓Information accuracy for multi-agent coordination
	[18]	WiFi, WiMAX, ZigBee	-	✓Algorithm of control and energy management for MG	MATLAB	✓Real-Time control of MG
	[78]	-	-	✓Power and control architecture	-	✓Increase RER penetration in MG
	[79]	WiFi	Modbus	✓Time delay, Transmission error rate, Coverage	Laboratory Testbed	✓Energy management, Monitoring, and control
Communication Technology-based	[80]	ZigBee	-	✓Defining required data to be transferred and a suitable coding	PSCAD	✓Data management scheme to overcome Low data transfer rate of ZigBee
	[52]	LoRaWAN	-	✓Time-on-air, duty cycle, and packet delivery ratio	field test	✓Provide multihop solution for P2P (Peer to Peer) communicate between LoRa devices to prove LoRa as communication method in regional MG
	[57]	5G	OPC UA	-	Functional Mockup Interface (FMI) for modeling prosumers	✓multi-MG EMS

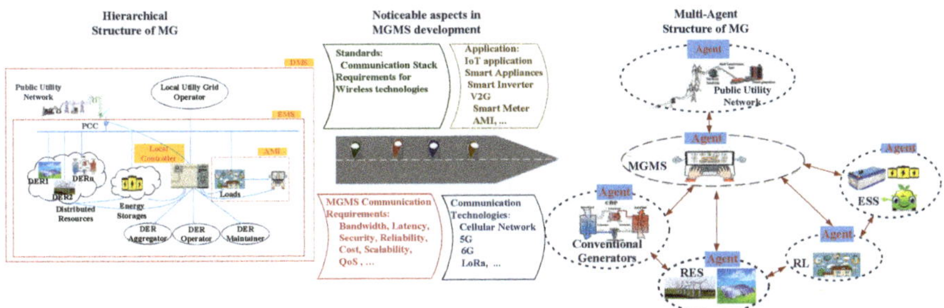

Figure 4. Basis of MGMS development.

5.3. MGMS IoT Integrated

MGMS was also revolutionized by the high penetration of IoT devices, especially in HEMS, which motivates the effective contribution of MG in DR through AMI and smart meters. This enhancement provides an infrastructure to schedule each home appliance inside MG to participate in DR resulting in a rocketing number of nodes. 5G technology based on MIMO characteristics can support communication with this number of nodes. LoRaWAN according to its lower power consumption, high data rate, and secure communication is another appropriate candidate for smart monitoring [81,82].

Figure 4 depicts an effective roadmap for applying wireless technology elements in MGMS, including communication specification, application, standards, and new technologies' requirements of MGMS.

6. Future Work

During an evolution towards MAS and multi-MG, emerging system interconnection has increased the operation and maintenance of communication infrastructure. Wireless communication enhancement with the introduction of new approaches such as 5G has been providing a solution to tackle this problematic issue by conservation cost of the system through network slicing and cloud concepts. Future work will conduct a determination of MGMS service-level requirements mapping into a service-driven 5G network and in a holistic approach for cloud-based communication structures for multi-MG and MAS investigation with wireless communication and IoT assistance.

7. Conclusions

MG is a solution, which facilitates SG implementation by using RER along with the control and management of loads and sources, which overcomes the intermittent nature of RER. Therefore, the management of MG is a key element that requires a robust communication infrastructure. This paper also revealed the role of MG in the power system by racketing its penetration changes to active elements, which offers AS to a utility public network. Because of its simplicity, affordability, and superb performance of wireless communication, especially in remote places, we noticed communication technologies and their characteristics in this paper in a way that summarized and reviewed applying them in different levels of MGMS as a prominent contribution of this paper, which has been ignored in recent studies. According to the MG diversity of applications, there are different control and management scenarios. Investigation of a holistic wireless employment roadmap in MGMS was another achievement of this study, which covers different control methodology scenarios and correspondence of new enhancements in wireless technology characteristics such as 5G and LoRa with new requirements of MG as an active element of the power system. This roadmap highlighted wireless enhancement application in the implementation of the developed structure of MGMS, including MAS, and higher-aspect multi-MG structure of SG. Mapping service-level requirements of these developed

MGMS to network slicing and cloud concepts through application of recent wireless technologies introduced as a new horizon of this research area.

Author Contributions: Conceptualization, L.T. and H.Y.; Data curation, L.T. and H.Y.; Writing—original draft preparation, L.T., H.Y. and M.J.P.; Writing—review and editing, L.T., H.Y. and M.J.P.; Supervision, L.T. and H.Y. All authors have read and agreed to the published version of the manuscript.

Funding: This work was supported by the National Research Foundation of Korea (NRF) grant funded by the Korea government (MSIT) (No. 2019M3F2A1073179) and by Korea Electric Power Corporation (Grant number: R17XA05-66).

Conflicts of Interest: The authors declare no conflict of interest.

Abbreviations

The following abbreviations are used in this manuscript:

AMI	Advanced Metering Interface
AS	Ancillary Services
BBPLC	Broad Band PLC
CSS	Chirp Spread Spectrum
CT	Current Transformer
DER	Distributed Energy Resources
DMS	Distribution Management System
DR	Demand Response
DSO	Distribution System Operator
DSSS	Direct Sequence Spread Spectrum
eMBB	Enhanced Mobile Broadband
EMS	Energy Management System
ESS	Energy Storage System
EV	Electric Vehicle
5G	Fifth-Generation
FMI	Functional Mockup Interface
FSK	Frequency Shifting Key
GEO	Geostationary Earth Orbits
GOOSE	Generic Object-Oriented Substation Event
GSM	Global System for Mobile
FAN	Field Area Network
HAN	Home Area Network
HD	High-Definition
HEMS	Home Energy Management System
IEC	International Electrotechnical Commission
IED	Intelligent Electronic Devices
IEEE	Institute of Electrical and Electronics Engineers
IoT	Internet of Things
ISM	Industrial, Scientific, and Medical
ITU	International Telecommunication Union
LAN	Local Area Network
LEO	Low Earth Orbits
LPWAN	Low Power WAN
LTE	Long-Term Evolution
MAS	Multi-Agent System
MEO	Medium Earth Orbit
MG	Micro-Grid
MGMS	Micro-Grid Management System
MIMO	Multiple-Input Multiple-Output
MMS	Manufacturing Message Specification
mMTC	Massive Machine Type Communications
mm-Wave	Millimeter-Wave
NBPLC	Narrow Band PLC
OSI	Open System Interaction
PCC	Point of Common Coupling
PLC	Power Line Carrier
PMU	Phasor Measurements Units
P2P	Peer to Peer

PT	Potential Transformer
QoS	Quality of Services
RER	Renewable Energy Resources
RL	Responsive Loads
SG	Smart Grid
SGIRM	Smart Grid Interoperability Reference Model
SV	Sampled Value
TSO	Transmission System Operator
UHF	Ultra-High Frequency
UNBPLC	Ultra-Narrow Band PLC
uRLLC	Ultra-reliable and Low-latency Communications
V2G	Vehicle-to-Grid
VR	Virtual Reality
WAN	Wide-Area Network
WLAN	Wireless LAN
WPAN	Wireless Personal Area Network
XMPP	eXtensible Messaging and Presence Protocol

References

1. Vera, G.; Yimy, E.; Dufo-López, R.; Bernal-Agustín, J.L. Energy management in microgrids with renewable energy sources: A literature review. *Appl. Sci.* **2019**, *9*, 3854. [CrossRef]
2. Lasseter, R.H. Smart Distribution: Coupled Microgrids. *Proc. IEEE* **2011**, *99*, 1074–1082. [CrossRef]
3. *IEEE Standard for the Specification of Microgrid Controllers*; IEEE Std 2030.7; IEEE Standards Association: Piscataway, NJ, USA, 2018.
4. Piagi, P.; Lasseter, R.H. Autonomous control of microgrids. In Proceedings of the IEEE Power Engineering Society General Meeting, Montreal, QC, Canada, 18–22 June 2006; p. 8.
5. Jiang, Z.; Yu, X. Power electronics interfaces for hybrid DC and AC-linked microgrids. In Proceedings of the IEEE 6th International Power Electronics and Motion Control Conference, Wuhan, China, 17–20 May 2009; pp. 730–736.
6. Wu, D.; Dragicevic, T.; Vasquez, J.C.; Guerrero, J.M.; Guan, Y. Secondary coordinated control of islanded microgrids based on consensus algorithms. In Proceedings of the IEEE Energy Conversion Congress and Exposition (ECCE), Pittsburgh, PA, USA, 14–18 September 2014; pp. 4290–4297.
7. Han, Y.; Zhang, K.; Li, H.; Coelho, E.A.A.; Guerrero, J.M. MAS-based distributed coordinated control and optimization in microgrid and microgrid clusters: A comprehensive overview. *IEEE Trans. Power Electron.* **2017**, *33*, 6488–6508. [CrossRef]
8. Zaidi, A.A.; Kupzog, F. Microgrid automation-a self-configuring approach. In Proceedings of the IEEE International Multitopic Conference, Karachi, Pakistan, 23–24 December 2008; pp. 565–570.
9. Jiang, T.; Costa, L.M.; Siebert, N.; Tordjman, P. Automated microgrid control systems. *IET Cired Open Access Proc. J.* **2017**, 961–964. [CrossRef]
10. Venkata, S.M.; Shahidehpour, M. Microgrid controllers: The brain, heart, & soul of microgrid automation. *IEEE Power Energy Mag.* **2017**, *15*, 16–22.
11. Ustun, T.S.; Ozansoy, C.; Zayegh, A. Simulation of communication infrastructure of a centralized microgrid protection system based on IEC 61850-7-420. In Proceedings of the IEEE Third International Conference on Smart Grid Communications (SmartGridComm), Tainan, Taiwan, 5–8 November 2012; pp. 492–497.
12. Mao, M.; Mei, F.; Jin, P.; Chang, L. Application of IEC61850 in energy management system for microgrids. In Proceedings of the IEEE 5th International Symposium on Power Electronics for Distributed Generation Systems (PEDG), Galway, Ireland, 24–27 June 2014; pp. 1–5.
13. Yoo, B.; Yang, H.S.; Yang, S.; Jeong, Y.S.; Kim, W.Y. CAN to IEC 61850 for Microgrid system. In Proceedings of the IEEE International Conference on Advanced Power System Automation and Protection, Beijing, China, 16–20 October 2011; Volume 2, pp. 1219–1224.
14. Zaheeruddin; Manas, M. Renewable energy management through microgrid central controller design: An approach to integrate solar, wind and biomass with battery. *Elsevier Energy Rep.* **2015**, *1*, 156–163.
15. Palizban, O.; Kauhaniemi, K.; Guerrero, J.M. Microgrids in active network management–part II: System operation, power quality and protection. *Elsevier Renew. Sustain. Energy Rev.* **2014**, *36*, 440–451. [CrossRef]

16. Chavan, M.P.; Devi, R.J. Survey of Communication System for DG's and Microgrid in Electrical Power Grid. *Int. Res. J. Eng. Technol.* **2016**, *3*, 1155–1164.

17. Islam, M.; Lee, H.H. Microgrid communication network with combined technology. In Proceedings of the IEEE 5th International Conference on Informatics, Electronics and Vision (ICIEV), Dhaka, Bangladesh, 13–14 May 2016; pp. 423–427.

18. Elkhorchani, H.; Grayaa, K. Smart micro Grid power with wireless communication architecture. In Proceedings of the IEEE International Conference on Electrical Sciences and Technologies in Maghreb (CISTEM), Tunis, Tunisia, 3–6 November 2014; pp. 1–10.

19. Weimer, J.; Xu, Y.; Fischione, C.; Johansson, K.H.; Ljungberg, P.; Donovan, C.; Sutor, A.; Fahlén, L.E. A virtual laboratory for micro-grid information and communication infrastructures. In Proceedings of the 3rd IEEE PES Innovative Smart Grid Technologies Europe (ISGT Europe), Berlin, Germany, 14–17 October 2012; pp. 1–6.

20. Marzal, S.; Salas, R.; González-Medina, R.; Garcerá, G.; Figueres, E. Current challenges and future trends in the field of communication architectures for microgrids. *Elsevier Renew. Sustain. Energy Rev.* **2018**, *82*, 3610–3622. [CrossRef]

21. Safdar, S.; Hamdaoui, B.; Cotilla-Sanchez, E.; Guizani, M. A survey on communication infrastructure for micro-grids. In Proceedings of the IEEE 9th International Wireless Communications and Mobile Computing Conference (IWCMC), Sardinia, Italy, 1–5 July 2013; pp. 545–550.

22. Mariam, L.; Basu, M.; Conlon, M.F. A review of existing microgrid architectures. *Hindawi J. Eng.* **2013**, *2013*, 937614. [CrossRef]

23. Bani-Ahmed, A.; Weber, L.; Nasiri, A.; Hosseini, H. Microgrid communications: State of the art and future trends. In Proceedings of the IEEE 2014 International Conference on Renewable Energy Research and Application (ICRERA), Milwaukee, WI, USA, 19–22 October 2014; pp. 780–785.

24. Mavrokefalidis, C.; Ampeliotis, D.; Berberidis, K. A study of the communication needs in micro-grid systems. In Proceedings of the XXXIInd General Assembly and Scientific Symposium of the International Union of Radio Science (URSI GASS), Montreal, QC, Canada, 19–26 August 2017; pp. 1–4.

25. Mai, T.; Haque, A.; Vo, T.; Nguyen, P.; Pham, M.C. Development of ICT infrastructure for Physical LV Microgrids. In Proceedings of the IEEE International Conference on Environment and Electrical Engineering and 2018 IEEE Industrial and Commercial Power Systems Europe (EEEIC/I&CPS Europe), Palermo, Italy, 12–15 June 2018; pp. 1–6.

26. Arbab-Zavar, B.; Palacios-Garcia, E.J.; Vasquez, J.C.; Guerrero, J.M. Smart Inverters for Microgrid Applications: A Review. *Energies* **2019**, *12*, 840. [CrossRef]

27. Amicarelli, E.; Tran, Q.T.; Bacha, S. Multi-agent system for day-ahead energy management of microgrid. In Proceedings of the IEEE 2016 18th European Conference on Power Electronics and Applications (EPE'16 ECCE Europe), Karlsruhe, Germany, 5–9 September 2016; pp. 1–10.

28. Hossain, E.; Han, Z.; Poor, H.V. *Smart Grid Communications and Networking*; Cambridge University Press: Cambridge, UK, 2012.

29. Merabet, G.H.; Essaaidi, M.; Talei, H.; Abid, M.R.; Khalil, N.; Madkour, M.; Benhaddou, D. Applications of multi-agent systems in smart grids: A survey. In Proceedings of the IEEE 2014 International Conference on Multimedia Computing and Systems (ICMCS), Marrakech, Morocco, 14–16 April 2014; pp. 1088–1094.

30. Howell, S.; Rezgui, Y.; Hippolyte, J.L.; Jayan, B.; Li, H. Towards the next generation of smart grids: Semantic and holonic multi-agent management of distributed energy resources. *Elsevier Renew. Sustain. Energy Rev.* **2017**, *77*, 193–214. [CrossRef]

31. Priyadarshana, H.; Sandaru, M.K.; Hemapala, K.; Wijayapala, W. A review on Multi-Agent system based energy management systems for micro grids. *AIMS Energy* **2019**, *7*, 924. [CrossRef]

32. Karimi, H.; Jadid, S. Optimal energy management for multi-microgrid considering demand response programs: A stochastic multi-objective framework. *Energy* **2020**, *195*, 116992. [CrossRef]

33. Kuzlu, M.; Pipattanasomporn, M.; Rahman, S. Communication network requirements for major smart grid applications in HAN, NAN, and WAN. *Elsevier Comput. Netw.* **2014**, *67*, 74–88. [CrossRef]

34. *Basic Communication Structure–Distributed Energy Resources Logical Nodes*; IEC61850-7-420; IEC: Geneva, Switzerland, 2009.

35. *Specific Communication Service Mapping (SCSM)—Mapping to Extensible Messaging Presence Protocol (XMPP)*; IEC61850-8-2; IEC: Geneva, Switzerland, 2018.

36. *Object Models for Power Converters in Distributed Energy Resources (DER) Systems*; IEC61850-90-7; IEC: Geneva, Switzerland, 2013.

37. *Wide Area Network Engineering Guidelines*; IEC61850-90-12; IEC: Geneva, Switzerland, 2015.

38. *Communications for Monitoring and Control of Wind Power Plants—Information Models*; IEC 61400-25-2; IEC: Geneva, Switzerland, 2015.

39. *IEEE Guide for Monitoring, Information Exchange, and Control of Distributed Resources Interconnected with Electric Power Systems*; IEEE Std 1547.3-2007; IEEE Standards Association: Piscataway, NJ, USA, 2007.

40. *IEEE Guide for Smart Grid Interoperability of Energy Technology and Information Technology Operation with the Electric Power System (EPS), End-Use Applications, and Loads*; IEEE Std 2030-2011; IEEE Standards Association: Piscataway, NJ, USA, 2011.

41. Tightiz, L.; Yang, H. Survey of IEC61850 M2M Interface based on IoT Protocols in Smart Grid Environment. In Proceedings of the Korea Institute Of Communication Sciences (KICS) Winter Conference, YongPyung Resort, Korea, 23–25 February 2019; Volume 1; pp. 382–385.

42. Jindal, A.; Marnerides, A.K.; Gouglidis, A.; Mauthe, A.; Hutchison, D. Communication standards for distributed renewable energy sources integration in future electricity distribution networks. In Proceedings of the IEEE International Conference on Acoustics, Speech and Signal Processing (ICASSP), Brighton, UK, 12–17 May 2019; pp. 8390–8393.

43. Sato, T.; Kammen, D.M.; Duan, B.; Macuha, M.; Zhou, Z.; Wu, J.; Tariq, M.; Asfaw, S.A. *Smart Grid Standards: Specifications, Requirements, and Technologies*; John Wiley & Sons: Hoboken, NJ, USA, 2015.

44. Basso, T.; DeBlasio, R. *IEEE Smart Grid Series of Standards IEEE 2030 (Interoperability) and IEEE 1547 (Interconnection) Status*; Technical Report; National Renewable Energy Lab. (NREL): Golden, CO, USA, 2012.

45. Gungor, V.C.; Sahin, D.; Kocak, T.; Ergut, S.; Buccella, C.; Cecati, C.; Hancke, G.P. Smart grid technologies: Communication technologies and standards. *IEEE Trans. Ind. Inform.* **2011**, *7*, 529–539. [CrossRef]

46. Bouhafs, F.; Mackay, M.; Merabti, M. *Communication Challenges and Solutions in the Smart Grid*; Springer: Berlin, Germany, 2014.

47. Raza, N.; Akbar, M.Q.; Aized Amin Soofi, S.A. Study of Smart Grid Communication Network Architectures and Technologies. *Sci. Res. Publ. J. Comput. Commun.* **2019**, *7*, 19–29. [CrossRef]

48. Ghorbanian, M.; Dolatabadi, S.H.; Masjedi, M.; Siano, P. Communication in Smart Grids: A Comprehensive Review on the Existing and Future Communication and Information Infrastructures. *IEEE Syst. J.* **2019**, *13*, 4001–4014. [CrossRef]

49. Saleem, Y.; Crespi, N.; Rehmani, M.H.; Copeland, R. Internet of Things-Aided Smart Grid: Technologies, Architectures, Applications, Prototypes, and Future Research Directions. *IEEE Access* **2019**, *7*, 62962–63003. [CrossRef]

50. Hui, H.; Ding, Y.; Shi, Q.; Li, F.; Song, Y.; Yan, J. 5G network-based Internet of Things for demand response in smart grid: A survey on application potential. *Appl. Energy* **2020**, *257*, 113972. [CrossRef]

51. Mikhaylov, K.; Petäjäjärvi, J.; Haapola, J.; Pouttu, A. D2D communications in lorawan low power wide area network: From idea to empirical validation. In Proceedings of the 2017 IEEE International Conference on Communications Workshops (ICC Workshops), Kansas City, MO, USA, 20–24 May 2017; pp. 737–742.

52. Zhou, W.; Tong, Z.; Dong, Z.Y.; Wang, Y. LoRa-Hybrid: A LoRaWAN Based multihop solution for regional microgrid. In Proceedings of the 2019 IEEE 4th International Conference on Computer and Communication Systems (ICCCS), Singapore, 23–25 February 2019; pp. 650–654.

53. Abbasi, M.; Khorasanian, S.; Yaghmaee, M.H. Low-Power Wide Area Network (LPWAN) for Smart grid: An in-depth study on LoRaWAN. In Proceedings of the 2019 5th Conference on Knowledge Based Engineering and Innovation (KBEI), Tehran, Iran, 28 February–1 March 2019; pp. 22–29.

54. Al-Turjman, F.; Abujubbeh, M. IoT-enabled smart grid via SM: An overview. *Future Gener. Comput. Syst.* **2019**, *96*, 579–590. [CrossRef]

55. Kabalci, Y. A survey on smart metering and smart grid communication. *Renew. Sustain. Energy Rev.* **2016**, *57*, 302–318. [CrossRef]

56. Sohraby, K.; Minoli, D.; Occhiogrosso, B.; Wang, W. A review of wireless and satellite-based M2M/IoT services in support of smart grids. *Mob. Netw. Appl.* **2018**, *23*, 881–895. [CrossRef]

57. Gross, S.; Ponci, F.; Monti, A. Multi-Microgrid Energy Management System in Times of 5G. In Proceedings of the 2019 IEEE International Conference on Communications, Control, and Computing Technologies for Smart Grids (SmartGridComm), Beijing, China, 21–23 October 2019; pp. 1–6.

58. Digra, R.K.; Pandey, R.K. Multi-agent control coordination of Microgrid. In Proceedings of the 2013 Students Conference on Engineering and Systems (SCES), Allahabad, India, 12–14 April 2013; pp. 1–5.

59. Dou, C.; Liu, B. Multi-Agent Based Hierarchical Hybrid Control for Smart Microgrid. *IEEE Trans. Smart Grid* **2013**, *4*, 771–778. [CrossRef]

60. Morstyn, T.; Hredzak, B.; Agelidis, V.G. Control Strategies for Microgrids With Distributed Energy Storage Systems: An Overview. *IEEE Trans. Smart Grid* **2018**, *9*, 3652–3666. [CrossRef]

61. Hasanuzzaman Shawon, M.; Muyeen, S.M.; Ghosh, A.; Islam, S.M.; Baptista, M.S. Multi-Agent Systems in ICT Enabled Smart Grid: A Status Update on Technology Framework and Applications. *IEEE Access* **2019**, *7*, 97959–97973. [CrossRef]

62. Bag, G.; Thrybom, L.; Hovila, P. Challenges and opportunities of 5G in power grids. *IET CIRED Open Access Proc. J.* **2017**, *2017*, 2145–2148. [CrossRef]

63. Zia, M.F.; Elbouchikhi, E.; Benbouzid, M. Microgrids energy management systems: A critical review on methods, solutions, and prospects. *Appl. Energy* **2018**, *222*, 1033–1055. [CrossRef]

64. Kim, Y.J.; Wang, J.; Lu, X. A framework for load service restoration using dynamic change in boundaries of advanced microgrids with synchronous-machine DGs. *Trans. Smart Grid* **2016**, *9*, 3676–3690. [CrossRef]

65. Madureira, A.; Lopes, J.P. Ancillary services market framework for voltage control in distribution networks with microgrids. *Electr. Power Syst. Res.* **2012**, *86*, 1–7. [CrossRef]

66. Wu, Y.; Hu, W.; Yang, Z.; Jing, J.; Wu, J. Service Restoration of Active Distribution Network Considering the Islanded Operation of Distributed Generation and Micro-grid. In Proceedings of the 2018 IEEE 2nd International Electrical and Energy Conference (CIEEC), Beijing, China, 4–6 November 2018; pp. 309–313.

67. Gheisarnejad, M.; Khooban, M.H.; Dragicevic, T. The future 5G network based secondary load frequency control in maritime microgrids. *IEEE J. Emerg. Sel. Top. Power Electron.* **2019**, *8*, 836–844. [CrossRef]

68. Covatti, F.; Winter, J.M.; Muller, I.; Pereira, C.E.; Netto, J.C. Wireless communication for IEC61850: A WirelessHART gateway proposal. In Proceedings of the IEEE International Conference on Industrial Technology (ICIT), Busan, Korea, 26 February–1 March 2014; pp. 754–759.

69. Kanabar, P.M.; Kanabar, M.G.; El-Khattam, W.; Sidhu, T.S.; Shami, A. Evaluation of communication technologies for IEC 61850 based distribution automation system with distributed energy resources. In Proceedings of the 2009 IEEE Power & Energy Society General Meeting, Calgary, AB, Canada, 26–30 July 2009; pp. 1–8.

70. Rinaldi, S.; Ferrari, P.; Ali, N.M.; Gringoli, F. IEC 61850 for micro grid automation over heterogeneous network: Requirements and real case deployment. In Proceedings of the IEEE 13th International Conference on Industrial Informatics (INDIN), Cambridge, UK, 22–24 July 2015; pp. 923–930.

71. Yang, X.; Wang, Y.; Zhang, Y.; Xu, D. Modeling and Analysis of Communication Network in Smart Microgrids. In Proceedings of the 2nd IEEE Conference on Energy Internet and Energy System Integration (EI2), Beijing, China, 20–22 October 2018; pp. 1–6.

72. Hussain, S.M.S.; Tak, A.; Ustun, T.S.; Ali, I. Communication Modeling of Solar Home System and Smart Meter in Smart Grids. *IEEE Access* **2018**, *6*, 16985–16996. [CrossRef]

73. Nagothu, K.; Kelley, B.; Jamshidi, M.; Rajaee, A. Persistent Net-AMI for Microgrid Infrastructure Using Cognitive Radio on Cloud Data Centers. *IEEE Syst. J.* **2012**, *6*, 4–15. [CrossRef]

74. Ustun, T.S.; Khan, R.H. Multiterminal hybrid protection of microgrids over wireless communications network. *IEEE Trans. Smart Grid* **2015**, *6*, 2493–2500. [CrossRef]

75. Ci, S.; Qian, J.; Wu, D.; Keyhani, A. Impact of wireless communication delay on load sharing among distributed generation systems through smart microgrids. *IEEE Wirel. Commun.* **2012**, *19*, 24–29.

76. Oureilidis, K.O.; Demoulias, C.S. Microgrid wireless energy management with energy storage system. In Proceedings of the 47th International Universities Power Engineering Conference (UPEC), London, UK, 4–7 September 2012; pp. 1–6.

77. Liang, H.; Choi, B.J.; Zhuang, W.; Shen, X.; Awad, A.S.A.; Abdr, A. Multiagent coordination in microgrids via wireless networks. *IEEE Wirel. Commun.* **2012**, *19*, 14–22. [CrossRef]

78. Kwasinski, A.; Kwasinski, A. Operational aspects and power architecture design for a microgrid to increase the use of renewable energy in wireless communication networks. In Proceedings of the International Power Electronics Conference (IPEC-Hiroshima 2014 - ECCE ASIA), Hiroshima, Japan, 18–21 May 2014; pp. 2649–2655.

79. Siow, L.K.; So, P.L.; Gooi, H.B.; Luo, F.L.; Gajanayake, C.J.; Vo, Q.N. Wi-Fi based server in microgrid energy management system. In Proceedings of the TENCON IEEE Region 10 Conference, Singapore, 23–26 November 2009; pp. 1–5.
80. Setiawan, M.A.; Shahnia, F.; Rajakaruna, S.; Ghosh, A. ZigBee-Based Communication System for Data Transfer Within Future Microgrids. *IEEE Trans. Smart Grid* **2015**, *6*, 2343–2355. [CrossRef]
81. Zeinali, M.; Thompson, J.; Khirallah, C.; Gupta, N. Evolution of home energy management and smart metering communications towards 5G. In Proceedings of the 2017 8th International Conference on the Network of the Future (NOF), London, UK, 22–24 November 2017; pp. 85–90.
82. Dragičević, T.; Siano, P.; Prabaharan, S. Future generation 5G wireless networks for smart grid: A comprehensive review. *Energies* **2019**, *12*, 2140.

Article

Distributed Hierarchical Consensus-Based Economic Dispatch for Isolated AC/DC Hybrid Microgrid

Ke Jiang, Feng Wu *, Linjun Shi and Keman Lin

College of Energy and Electrical Engineering, Hohai University, Nanjing 211100, China;
jiaohuang051@163.com (K.J.); eec@hhu.edu.cn (L.S.); linkeman@hhu.edu.cn (K.L.)
* Correspondence: wufeng@hhu.edu.cn; Tel.: +86-153-0518-7449

Received: 12 May 2020; Accepted: 16 June 2020; Published: 20 June 2020

Abstract: In this paper, a distributed hierarchical consensus algorithm is proposed to solve the economic dispatch (ED) problem for the isolated AC/DC hybrid microgrid. At first, the whole nodes of the AC/DC hybrid microgrid are divided into two parts, that is, the leadership layer nodes and the tracking layer nodes. The leadership layer nodes update the data through their own feedback elements, while the tracking layer nodes receive the information from the leadership layer nodes and update the data. After several iterations, the two different layer nodes obtain the same state, which realizes the dynamic active power balance of the whole AC/DC microgrid. Besides, the AC sub-grid and DC sub-grid can also realize the power balance by the proposed algorithm, and the energy storage units will absorb or release active power to meet the power demand in the respective section. In addition, the constraints of the nodes are also taken into account to guarantee that the power nodes in the AC/DC hybrid microgrid should operate within their own limitations, which is necessary to realize the ED for the considered hybrid microgrid. Finally, case study and simulation results are provided to illustrate the effectiveness of the proposed hierarchal method.

Keywords: isolated AC/DC hybrid microgrid; two-layer consensus method; dynamic economic dispatch; distributed hierarchical strategy

---

1. Introduction

As an effective way of integrating distributed power generations and load, microgrids have obtained more and more attention as applications [1–4]. Due to the microgrid containing various distributed power sources, such as photovoltaic (PV), wind turbine (WT), and so on, the distributed power sources have their own different characteristics. The traditional control methods will be invalid to be used to solve the control problem of microgrids. Therefore, some new control strategies must be studied to deal with the microgrid control method. As of now, there are several results to realize the different control strategies in microgrids [5,6].

The traditional centralized control method needs a control center, through which the control order is sent. The receiving element obtains the signal sent from the control center and makes a process, then executes the relevant actions. However, with the large number of distributed power sources permeating into the microgrid, the traditional centralized control strategies may be ineffective to realize satisfactory results. On the other hand, once the control center is out of service, the whole system will be unable to work, which means an enormous loss. The distributed control method can easily avoid the above problem, the main reason being that the distributed control strategy does not need the control center, it is just used for the information exchange between one unit and its neighbors, finally obtaining a whole system of information sharing. If one or more of the units are out of service, the rest of the units can continue to process the information exchange, finally resulting in complete information sharing. Therefore, compared with the traditional control methods, the distributed control strategies are superior [7,8]. In this paper, the distributed control method is utilized to solve the economic dispatch (ED) problem of the AC/DC hybrid microgrid.

On the other hand, with the rapid development of microgrid technology, the connection is closing between power grids, especially the relationship between the AC microgrid and the DC microgrid. Due to these two microgrids having different characteristics, the control methods and optimizing strategies are disparate. But the final goals are similar, that is, maintain the microgrid operating in a balanceable and safe state. The information exchange of the AC/DC hybrid microgrid is complicated and bidirectional, which is a challenging work.

In addition, the ED problem is significant for keeping the normal operation of the microgrid, and the rational dispatch is beneficial to reduce cost and increase profits. By power dispatching, the power demands between loads and generators are satisfied, and the stability of the power grid is enhanced. With the development of distributed control technology, the distributed consensus strategies are widely used to solve the ED problem. The proposed distributed method is also an exploration to realize the rational ED for the AC/DC microgrid.

2. Literature Review

It should be pointed out that the distributed consensus strategies have become promising technologies after decades of development. Up to now, there have been a large number of results about the applications of distributed control strategies. In Reference [9], by using the consensus algorithm, for an island microgrid, the optimal resource management strategy was presented. In Reference [10], the authors studied the power balance problem of the DC microgrid by using the consensus method. In References [11,12], the consensus algorithms were presented for solving the problems of the voltage stability and compensation in the microgrid, respectively. But the above references do not refer to the distributed hierarchical control thought. In Reference [13], for the DC microgrid, the distributed hierarchical control method was designed. A three-layer control framework for the static and dynamic performance of the microgrid was proposed in Reference [14]. In Reference [15], the distributed hierarchical control problem of the island microgrid was solved by proposing a network control scheme based on a robust communication algorithm. In Reference [16], by linearizing the input and output feedback, the second-order voltage problem in the microgrid was transformed into a second-order linear synchronization tracking problem, and a distributed hierarchical control strategy was developed. In Reference [17], a hierarchical control structure was developed for an autonomous microgrid, which could effectively adjust the photovoltaic output and realize the active balance in the considered microgrid. In Reference [18], for solving the problem of active regulation of the autonomous microgrid, the authors designed a two-stage controller and the optimal active power regulation controller. Although the above references are related to the hierarchical control thought, the object is simple microgrids, which does not involve the AC/DC hybrid microgrid.

With the development of research, many researchers have transferred their attention to the AC/DC hybrid microgrid. By using different study methods, more and more results about the AC/DC hybrid microgrid have been published [19–31]. In References [20,21], the voltage control problem was investigated by using different methods in the AC/DC hybrid microgrid. In References [22–24], the problems of power control and management of the AC/DC hybrid microgrid were discussed through different methods. For the interconnected microgrids, the authors designed an event-based consensus strategy for solving the coordinated power sharing problems in Reference [25]. In Reference [26], for solving the reactive power sharing and DC current sharing problem in the considered hybrid microgrids, a distributed coordination control method was developed. In References [27,28], two different bidirectional interlinking converters were studied for enhancing the transient performance of an islanded AC-DC hybrid microgrid. In Reference [29], an appropriate combinatorial optimization technique was proposed for solving the optimal sizing problem of AC-DC hybrid microgrids. In Reference [30], a straightforward and efficient method was studied to solve power flows' simultaneous problem of AC/DC hybrid microgrids.

On the other hand, as the basis for ensuring the economic and reliable operation of microgrids, the economic dispatch (ED) problem has also attracted special attention, and lots of results based on consensus algorithms have been developed [32,33]. In Reference [34], for solving the ED problem, the authors proposed a consensus-based algorithm for learning about the mismatch. In Reference [35], for the islanded microgrids, a new consensus-based method was developed, where the incremental cost of every agent is taken as the consensus variable, so their proposed method could quickly converge to the optimal solutions. In Reference [36], for solving the ED issue, an improved distributed consensus algorithm was proposed, where the feedback gains of the algorithm could be different and time varying. In Reference [37], considering the delay effect in the optimal operation of microgrids, the ED problem was discussed based on the consensus algorithm. In Reference [38], in the framework of hierarchical distributed theory, considering the power limitation of the DC microgrid, a constrained economic dispatch method was proposed. As for the ED problem for the AC/DC hybrid microgrid, some results have been developed. In Reference [39], for the considered hybrid microgrid, by considering the properties of AC and DC subsystems and the deficiencies in concentrated optimization, a management system based on consensus and coordination optimization was developed for the safe and economic operation. For discussing the economic operation of the hybrid AC/DC microgrid, a distributed control framework was proposed in Reference [40]. In References [41,42], a distributed economic dispatch method was developed based on the so-called finite-step consensus method, which could reach to the optimal solutions in finite steps. However, up to now, there are few results discussing hierarchical distributed strategy for the ED problem of the AC/DC hybrid microgrid. By fully taking into account the characteristics of the AC sub-grid and the DC sub-grid, how to connect the relationship between them and propose a reasonable control algorithm is an important issue for solving the ED problem for the AC/DC hybrid microgrid, which is an interesting and challenging topic.

Based on the above analysis and discussions, in this paper, a distributed two-layer consensus algorithm is proposed to investigate the ED problem of the AC/DC hybrid microgrid in detail. At first, the whole nodes of the AC/DC hybrid microgrid will be divided into two parts, that is, the leadership layer nodes and the tracking layer nodes. Then, by using the proposed consensus strategy, the data of the leadership layer and the tracking layer are uniformly processed to achieve the same increment cost (IC) of each node, which means that the ED problem is realized. Finally, some case study results and discussions are presented to demonstrate the usefulness of the developed hierarchical strategy.

The remainder of the article is arranged as follows: The optimal scheduling model for the considered AC/DC hybrid microgrid is presented in Section 3. In Section 4, for solving the ED problem, the two-layer consensus algorithms are proposed with or without constraints, respectively. In Section 5, the case study is designed to demonstrate the usefulness of the proposed distributed method. In the end, our main conclusions are given in Section 6.

3. Optimal Scheduling Model for the AC/DC Hybrid Microgrid

3.1. Structure Information of the AC/DC Hybrid Microgrid

In this paper, we mainly focus on the isolated AC/DC hybrid microgrid. Figure 1 presents the structure information of the considered model in this paper, which mainly contains an AC sub-grid, a DC sub-grid, and a power conversion unit that is in-lined between sub-grids. The wind turbine (WT) is connected to the micro-grid AC bus, and the solar photovoltaic unit (PV) is connected to the micro-grid DC bus, while the energy storage unit (BS) and the AC/DC load unit (Load) are in both parts. The AC sub-grids and DC sub-grids are connected by an AC/DC bidirectional converter (IC) between the AC bus and the DC bus, and M1 is an AC/DC tie line power monitoring point. The dashed lines in Figure 1 indicate the communication links between the various units.

To promote the utilization efficiency of the renewable energy and ensure reliable power supply, the operation of the isolated AC/DC hybrid microgrid is able to be grouped into three levels: (1) The distributed generations in the AC sub-grid and DC sub-grid are used for realizing power self-balancing in the region, (2) through the commutation tie line, the AC sub-grid and DC sub-grid can realize inter-area power balance, and (3) the electric energy transfer and peak-shaving are realized by charging and discharging the energy storage unit.

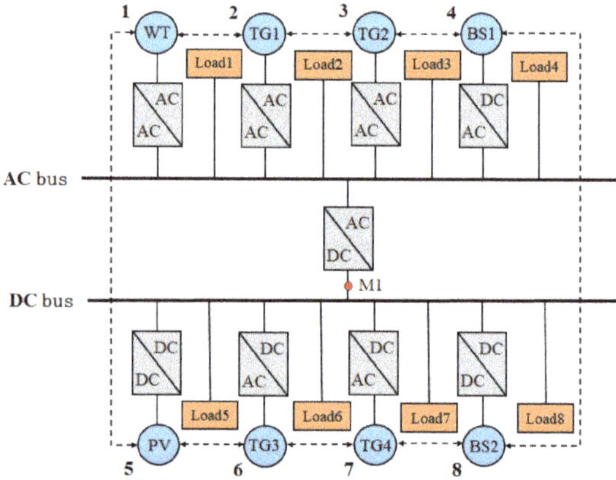

Figure 1. The structure of the considered AC/DC hybrid microgrid.

3.2. ED Model for the AC/DC Hybrid Microgrid

3.2.1. Objective Function

For the isolated microgrid, the dynamic economic dispatch strategy is based on the fact that all the distributed power supply equipment costs are fixed; meanwhile, the energy storage units can absorb or release power through their charging or discharging process. The objective function is formulated as

$$\min\left(\sum_{t=1}^{T}\left[\sum_{i\in S_{TG}} C_i(P_{TG,i}(t)) + \sum_{j\in S_{WT}} C_j\big(P_{WT,j}(t)\big) + \sum_{k\in S_{SL}} C_k\big(P_{SL,k}(t)\big) + \sum_{r\in S_{BS}} C_r(P_{BS,r}(t))\right]\right) \qquad (1)$$

where T is the number of time periods in the daily dispatching cycle. S_{TG} and S_{WT} are the sets of TG units and WT units, respectively. S_{SL} is the set of PV units, and S_{BS} represents the set of BS units. $P_{TG,i(t)}$, $P_{WT,j(t)}$, $P_{SL,k(t)}$, and $P_{BS,r(t)}$ are the active output power of the ith TG, jth WT, kth PV, and rth BS over a time period t, respectively. $C_i(P_{TG,i(t)})$, $C_j(P_{WT,j(t)})$, $C_k(P_{SL,k(t)})$, and $C_r(P_{BS,r(t)})$ are the generation cost/penalty functions of the corresponding unit set, respectively.

Traditional Generator Set Power Generation Cost Function

Without loss of generality, for a traditional generator set, the operating cost function can be modeled as the following quadratic function form [43]

$$C_i(P_{TG,i}(t)) = a_{TG,i}(P_{TG,i}(t))^2 + b_{TG,i}P_{TG,i}(t) + c_{TG,i}, \quad i \in S_{TG} \qquad (2)$$

where $a_{TG,i}$, $b_{TG,i}$, and $c_{TG,i}$ represent the cost function coefficients.

Wind Generator and Photovoltaic Unit Abandonment Wind Penalty Function

The wind and solar penalty functions are formulated as

$$C_j(P_{WT,j}(t)) = a_{WT,j}[P_{WT,j}(t) - P^{st}_{WT,j}(t)]^2 + b_{WT,j}[P_{WT,j}(t) - P^{st}_{WT,j}(t)] + c_{WT,j}, \quad j \in S_{WT} \qquad (3)$$

$$C_k(P_{SL,k}(t)) = a_{SL,k}[P_{SL,k}(t) - P^{st}_{SL,k}(t)]^2 + b_{SL,k}[P_{SL,k}(t) - P^{st}_{SL,k}(t)] + c_{SL,k}, \quad k \in S_{SL} \qquad (4)$$

where $a_{SL,k}, b_{SL,k}, c_{SL,k}$ and $a_{SL,k}, b_{SL,k}, c_{SL,k}$ are the cost coefficients of abandoned wind and light respectively, and $P_{WT,j}^{st}(t)$ and $P_{SL,k}^{st}(t)$ are the maximum value of the adjustable power of the jth WT and the kth PV at the moment t, respectively.

Energy Storage Unit Operating Cost Function

The operating cost function for the energy storage unit is represented as an over-origin quadratic function with an opening up of

$$C_r(P_{BS,r}(t)) = a_{BS,r}(P_{BS,r}(t))^2 + b_{BS,r}P_{BS,r}(t) + c_{BS,r}, \quad r \in S_{BS} \tag{5}$$

where $a_{BS,r}, b_{BS,r}, c_{BS,r}$ denotes the coefficient of cost function.

3.2.2. Constraint Conditions

For the safe and stable operation, some operational constraints, for instance, the supply and demand balance constraints of each unit, should be taken into account.

Constraints for Active Supply and Demand Balance

The supply–demand balance constraint for the considered hybrid microgrid is formulated as

$$\sum_{i \in S_{TG}} P_{TG,i}(t) + \sum_{j \in S_{WT}} P_{WT,j}(t) + \sum_{k \in S_{SL}} P_{SL,k}(t) + \sum_{r \in S_{BS}} P_{BS,r}(t) = \sum_{s \in S_{DM}} P_{DM,s}(t) \tag{6}$$

where S_{DM} is the set of all load cells in the AC/DC hybrid microgrid, and $P_{DM,s}(t)$ is the load demand value of load unit s at the moment t. Besides, the DC sub-grid and AC sub-grid of the microgrid must also meet the constraints of the active balance equation.

a. Constraint for DC side active balance:

$$\sum_{k \in S_{SL}} P_{SL,k}(t) + \sum_{r \in S_{BS}} P_{BS,r}(t) + P_{AC-DC}(t) = \sum_{s \in S_{DCDM}} P_{DCDM,s}(t) \tag{7}$$

where $P_{AC-DC}(t)$ is the interaction power between the AC and DC sub-grids through the commutation line and S_{DCDM} is the set of DC load cells.

b. Constraints for AC side active balance:

$$\sum_{i \in S_{TG}} P_{TG,i}(t) + \sum_{j \in S_{WT}} P_{WT,j}(t) - P_{AC-DC}(t) = \sum_{s \in S_{ACDM}} P_{ACDM,s}(t) \tag{8}$$

where S_{ACDM} is the set of AC load cells, and $P_{ACDM,s}(t)$ is the load value of DC load unit s at moment t.

Constraints for Traditional Generator Set Operation

a. Active power upper and lower bound:

$$P_{TG,i}^{min}(t) \leq P_{TG,i}(t) \leq P_{TG,i}^{max}(t) \tag{9}$$

where $P_{TG,i}^{min}(t)$ and $P_{TG,i}^{max}(t)$ are the active adjustable lower limit and upper limit of the conventional generator set at the moment t, respectively.

b. Constraint for output climbing:

$$-\Delta P_{TG,i}^{d} \leq P_{TG,i}(t) - P_{TG,i}(t-1) \leq \Delta P_{TG,i}^{u} \tag{10}$$

where $\Delta P_{TG,i}^{u}$ and $\Delta P_{TG,i}^{d}$ are the maximum of active power during the time period $[t-1, t]$ of the traditional generator set i.

Constraints for Operating the Energy Storage Unit

a. Charging and discharging power upper and lower bounds:

$$\underline{P}_{BS,r} \leq P_{BS,r}(t) \leq \overline{P}_{BS,r} \tag{11}$$

where $P_{BS,r}(t)$ is the output power of the rth BS at moment t, it is positive at the time of discharge and negative at the time of charging. $\overline{P}_{BS,r}$ and $\underline{P}_{BS,r}$ are the upper and lower bounds of the charging and discharging power of the rth BS, respectively.

b. Energy storage unit state of charge constraints:

$$SOC_{BS,r}^{\min} \leq SOC_{BS,r}(t) \leq SOC_{BS,r}^{\max} \tag{12}$$

where $SOC_{BS,r}^{\max}$ and $SOC_{BS,r}^{\min}$ are the upper and lower constraints the rth BS, respectively.

c. Energy storage unit capacity continuity constraints:

The relationship between the value of the energy storage unit at the moment and the previous moment can be expressed as

$$SOC_{BS,r}(t) = \begin{cases} SOC_{BS,r}(t-1) - \dfrac{P_{BS,r}(t)\eta_r^{ch}\Delta T}{E_r} & P_{BS,r}(t) < 0 \\[3mm] SOC_{BS,r}(t-1) - \dfrac{P_{BS,r}(t)\Delta T}{E_r\eta_r^{dis}} & P_{BS,r}(t) \geq 0 \end{cases} \tag{13}$$

where η_r^{ch} and η_r^{dis} are the charging and discharging efficiencies of the rth BS respectively, E_r represents the maximal capacity bound of the rth BS, and ΔT is the time interval from time $t-1$ to time t.

When considering the constraint for power capacity of the energy storage unit, the upper and lower limits of the active energy for energy storage unit are shown as

$$\begin{cases} P_{BS,r}^{\max}(t) = \min(\overline{P}_{BS,r}, P_{BS,r}^{ch}(t)) \\[2mm] P_{BS,r}^{\min}(t) = \max(\underline{P}_{BS,r}, P_{BS,r}^{dis}(t)) \end{cases} \tag{14}$$

where $P_{BS,r}^{\max}(t)$ and $P_{BS,r}^{\min}(t)$ are the power required by the rth BS to charge to the upper limit and discharge to the lower limit during the time period $[t-1,t]$, respectively. $P_{BS,r}^{\max}(t)$ and $P_{BS,r}^{\min}(t)$ denote the upper and lower limits of the rth BS at the moment t, respectively.

Wind Generator Photovoltaic Output Constraints

The adjustable ranges of the active output of wind power and photovoltaic power generation units are expressed as

$$0 \leq P_{WT,j}(t) \leq P_{WT,j}^{st}(t) \tag{15}$$

$$0 \leq P_{SL,k}(t) \leq P_{SL,k}^{st}(t) \tag{16}$$

AC and DC Tie Line Constraints

The power constraints for the commutation connection line of AC and DC sub-grids at the moment t are given as

$$P_{AC\_DC}^{\min} \leq P_{AC\_DC}(t) \leq P_{AC\_DC}^{\max} \tag{17}$$

where $P_{AC\_DC}^{\max}$ and $P_{AC\_DC}^{\min}$ are the limit of the power transmitted connection line, and if $P_{AC\_DC}^{\min}$ is a negative value, it indicates the upper limit of the power delivered from the DC sub-grid to the AC sub-grid.

3.2.3. Solutions

In our paper, for processing the single period distributed economic dispatching model, the Lagrangian multiplier method is used. Let λ be the Lagrangian multiplier; at first, the inequality constraints are ignored, then the considered optimization problem is given as

$$
\begin{aligned}
\min L \;=\; & \sum_{i \in S_{TG}} C_i(P_{TG,i}) + \sum_{j \in S_{WT}} C_j(P_{WT,j}) + \sum_{k \in S_{SL}} C_k(P_{SL,k}) + \sum_{r \in S_{BS}} C_r(P_{BS,r}) \\
& + \lambda \left(\sum_{s \in S_{DM}} P_{DM,s} - \sum_{i \in S_{TG}} P_{TG,i} - \sum_{j \in S_{WT}} P_{WT,j} - \sum_{k \in S_{SL}} P_{SL,k} - \sum_{r \in S_{BS}} P_{BS,r} \right)
\end{aligned}
\tag{18}
$$

By applying the Karush-Kuhn-Tucker (KKT) first-order optimality condition, the partial derivative of the decision quantity and Lagrangian multiplier can be obtained:

$$
\begin{cases}
\dfrac{\partial L}{\partial P_{TG,i}} = \dfrac{\partial C_i(P_{TG,i})}{\partial P_{TG,i}} - \lambda = 2a_{TG,i}P_{TG,i} + b_{TG,i} - \lambda = 0 \\[2mm]
\dfrac{\partial L}{\partial P_{WT,j}} = \dfrac{\partial C_j(P_{WT,j})}{\partial P_{WT,j}} - \lambda = 2a_{WT,j}(P_{WT,j} - P_{WT,j}^{st}) - \lambda = 0 \\[2mm]
\dfrac{\partial L}{\partial P_{SL,k}} = \dfrac{\partial C_k(P_{SL,k})}{\partial P_{SL,k}} - \lambda = 2a_{SL,k}(P_{SL,k} - P_{SL,k}^{st}) - \lambda = 0 \\[2mm]
\dfrac{\partial L}{\partial P_{BS,r}} = \dfrac{\partial C_r(P_{BS,r})}{\partial P_{BS,r}} - \lambda = 2a_{BS,r}P_{BS,r} - \lambda = 0 \\[2mm]
\dfrac{\partial L}{\partial \lambda} = \sum_{s \in S_{DM}} P_{DM,s} - \sum_{i \in S_{TG}} P_{TG,i} - \sum_{j \in S_{WT}} P_{WT,j} - \sum_{k \in S_{SL}} P_{SL,k} - \sum_{r \in S_{BS}} P_{BS,r} = 0
\end{cases}
\tag{19}
$$

When the operating incremental costs are equal, the Lagrangian function, L, takes the minimum value and the resulting λ is the optimal incremental costs. The corresponding operating cost factors are integrated to obtain the following unified form:

$$
2\tilde{a}_i \tilde{P}_i + \tilde{b}_i - \lambda = 0, \qquad i \in S_{TG} \cup S_{WT} \cup S_{SL} \cup S_{BS}
\tag{20}
$$

that is

$$
\tilde{P}_i = \frac{\lambda - \tilde{b}_i}{2\tilde{a}_i}
\tag{21}
$$

where $\tilde{a}_i$ and $\tilde{b}_i$ are the operating cost coefficient of the unit i after integration, and $\tilde{P}_i$ is the active output of the unit i.

4. Two-Layer Consensus Strategy

Due to the fact that the distributed control method does not need to set the control center, which only exchanges information between adjacent nodes and shares information, at last, the state of each node can reach consensus. Different from the traditional centralized control strategy, the distributed methods are able to effectively avoid some disadvantages caused by the failure of the control center, so they are more efficient and flexible.

In the microgrid, for simplicity, the power mismatch between the total generated power and total power demand is defined as

$$
\Delta P = \sum_{s \in S_{DM}} P_{DM,s}(t) - \sum_{i \in S_{TG}} P_{TG,i}(t) - \sum_{j \in S_{WT}} P_{WT,j}(t) - \sum_{k \in S_{SL}} P_{SL,k}(t) - \sum_{r \in S_{BS}} P_{BS,r}(t)
\tag{22}
$$

where the DC side and AC side power deviations can be expressed as

$$
\Delta P_{DC} = \sum_{s \in S_{DCDM}} P_{DCDM,s}(t) - \sum_{k \in S_{SL}} P_{SL,k}(t) - \sum_{r \in S_{BS}} P_{BS,r}(t) - P_{AC-DC}(t)
\tag{23}
$$

$$\Delta P_{AC} = \sum_{s \in S_{ACDM}} P_{ACDM,s}(t) - \sum_{i \in S_{TG}} P_{TG,i}(t) - \sum_{j \in S_{WT}} P_{WT,j}(t) + P_{AC-DC}(t) \tag{24}$$

4.1. Algorithm Design without Considering Constraints

For the considered hybrid microgrid, the nodes can be layered on DC and AC sub-grids. The leadership layer nodes include the feedback elements, which are used to update the self-data. The tracking layer nodes do not contain feedback elements, which receive leadership information to realize the data update.

For the hybrid model, the update algorithm of the leadership layer node is formulated as

$$\begin{cases} \lambda_i(t+1) = \sum\limits_{j \in N_i} d_{ij}\lambda_i(t) + \varepsilon \Delta P(t) \\ \tilde{P}_i(t+1) = \dfrac{\lambda_i(t+1) - \tilde{b}_i}{2\tilde{a}_i} \\ \Delta P(t+1) = \sum\limits_{s \in S_{DM}} P_{DM,s}(t) - \sum\limits_{i \in S_{TG}} P_{TG,i}(t) - \sum\limits_{j \in S_{WT}} P_{WT,j}(t) \\ \qquad\qquad - \sum\limits_{k \in S_{SL}} P_{SL,k}(t) - \sum\limits_{r \in S_{BS}} P_{BS,r}(t) \end{cases} \tag{25}$$

The update algorithm for the tracking layer node is

$$\begin{cases} \lambda_i(t+1) = \sum\limits_{j \in N_i} d_{ij}\lambda_i(t) \\ \tilde{P}_i(t+1) = \dfrac{\lambda_i(t+1) - \tilde{b}_i}{2\tilde{a}_i} \\ \Delta P(t+1) = \sum\limits_{s \in S_{DM}} P_{DM,s}(t) - \sum\limits_{i \in S_{TG}} P_{TG,i}(t) - \sum\limits_{j \in S_{WT}} P_{WT,j}(t) \\ \qquad\qquad - \sum\limits_{k \in S_{SL}} P_{SL,k}(t) - \sum\limits_{r \in S_{BS}} P_{BS,r}(t) \end{cases} \tag{26}$$

where d_{ij} is the node correlation coefficient, which is $d_{ij} = 2/(N_i + N_j + \delta)$, N_i and N_j are the number of nodes that directly connected to the node i and j respectively, and δ is a small positive number.

Combining Equations (25) and (26), a two-layer consensus algorithm is presented

$$\begin{cases} \lambda(t+1) = D\lambda(t) + E\Delta P(t) \\ \tilde{P}(t+1) = \tilde{A}\lambda(t+1) + \tilde{B} \\ \Delta P(t+1) = \sum\limits_{s \in S_{DM}} P_{DM,s}(t) - \sum \tilde{P}(t+1) \end{cases} \tag{27}$$

where

$$\lambda(t+1) = [\lambda_{TG,1}(t+1), \cdots, \lambda_{WT,1}(t+1), \cdots, \lambda_{SL,1}(t+1), \cdots, \lambda_{BS,1}(t+1), \cdots]^T$$

$$\tilde{P}(t+1) = [P_{TG,1}(t+1), \cdots, P_{WT,1}(t+1), \cdots, P_{SL,1}(t+1), \cdots, P_{BS,1}(t+1), \cdots]^T$$

$$\tilde{A} = \mathrm{diag}\left(\left[\frac{1}{2a_{TG,1}}\right], \cdots, \left[\frac{1}{2a_{WT,1}}\right], \cdots, \left[\frac{1}{2a_{SL,1}}\right], \cdots, \left[\frac{1}{2a_{BS,1}}\right], \cdots\right)$$

$$\tilde{B} = \left[-\frac{b_{TG,1}}{2a_{TG,1}}, \cdots, -\frac{b_{WT,1}}{2a_{WT,1}}, \cdots, -\frac{b_{SL,1}}{2a_{SL,1}} \cdots, -\frac{b_{BS,1}}{2a_{BS,1}}, \cdots\right]^T$$

$$D = \begin{bmatrix} 1 - \sum\limits_{j=1}^{n} d_{1j} & \cdots & d_{1i} & \cdots & d_{1n} \\ \cdots & \cdots & \cdots & \cdots & \cdots \\ d_{i1} & \cdots & 1 - \sum\limits_{j=1}^{n} d_{ij} & \cdots & d_{in} \\ \cdots & \cdots & \cdots & \cdots & \cdots \\ d_{n1} & \cdots & d_{ni} & \cdots & 1 - \sum\limits_{j=1}^{n} d_{nj} \end{bmatrix}$$

E is defined as a column vector, scalar 0 and feedback coefficient ε are taken as the elements. When the node is set as the leader node, the corresponding value is chosen as ε, otherwise, the value is chosen as 0.

For the DC sub-grid, the leadership layer node update strategy can be expressed as

$$
\begin{cases}
\lambda_{DC,i}(t+1) = \sum\limits_{j\in N_i} d_{ij}\lambda_{DC,i}(t) + \varepsilon\Delta P_{DC}(t) \\
\tilde{P}_{DC,i}(t+1) = \dfrac{\lambda_{DC,i}(t+1) - \check{b}_{DC,i}}{2\check{a}_{DC,i}} \\
\Delta P_{DC}(t+1) = \sum\limits_{s\in S_{DCDM}} P_{DCDM,s}(t) - \sum\limits_{k\in S_{SL}} P_{SL,k}(t) \\
\qquad\qquad\quad - \sum\limits_{r\in S_{BS}} P_{BS,r}(t) - P_{AC-DC}(t)
\end{cases}
\tag{28}
$$

The update algorithm for the tracking layer node is

$$
\begin{cases}
\lambda_{DC,i}(t+1) = \sum\limits_{j\in N_i} d_{ij}\lambda_{DC,i}(t) \\
\tilde{P}_{DC,i}(t+1) = \dfrac{\lambda_{DC,i}(t+1) - \check{b}_{DC,i}}{2\check{a}_{DC,i}} \\
\Delta P_{DC}(t+1) = \sum\limits_{s\in S_{DCDM}} P_{DCDM,s}(t) - \sum\limits_{k\in S_{SL}} P_{SL,k}(t) \\
\qquad\qquad\quad - \sum\limits_{r\in S_{BS}} P_{BS,r}(t) - P_{AC-DC}(t)
\end{cases}
\tag{29}
$$

For the AC sub-grid, the leadership layer node update strategy can be expressed as

$$
\begin{cases}
\lambda_{AC,i}(t+1) = \sum\limits_{j\in N_i} d_{ij}\lambda_{AC,i}(t) + \varepsilon\Delta P_{AC}(t) \\
\tilde{P}_{AC,i}(t+1) = \dfrac{\lambda_{AC,i}(t+1) - \check{b}_{AC,i}}{2\check{a}_{AC,i}} \\
\Delta P_{AC}(t+1) = \sum\limits_{s\in S_{ACDM}} P_{ACDM,s}(t) - \sum\limits_{i\in S_{TG}} P_{TG,i}(t) \\
\qquad\qquad\quad - \sum\limits_{j\in S_{WT}} P_{WT,j}(t) + P_{AC-DC}(t)
\end{cases}
\tag{30}
$$

The update algorithm for the tracking layer node is

$$
\begin{cases}
\lambda_{AC,i}(t+1) = \sum\limits_{j\in N_i} d_{ij}\lambda_{AC,i}(t) \\
\tilde{P}_{AC,i}(t+1) = \dfrac{\lambda_{AC,i}(t+1) - \check{b}_{AC,i}}{2\check{a}_{AC,i}} \\
\Delta P_{AC}(t+1) = \sum\limits_{s\in S_{ACDM}} P_{ACDM,s}(t) - \sum\limits_{i\in S_{TG}} P_{TG,i}(t) \\
\qquad\qquad\quad - \sum\limits_{j\in S_{WT}} P_{WT,j}(t) + P_{AC-DC}(t)
\end{cases}
\tag{31}
$$

Equations (25) and (26), (28) and (29), and (30) and (31) are the proposed algorithms that can be applied to the AC/DC hybrid microgrid, the DC sub-grid, and the AC sub-grid, respectively. Therefore, the proposed two-layer consensus control method is able to effectively solve the ED issue. Next, we will comprehensively consider the influence of various constraints, and modify the proposed algorithms.

4.2. Algorithm Design with Considering Constraints

When considering the active output of the generator, the update equation of the generator node can be expressed as

$$
\begin{cases}
P_{TG,i}(t) = P_{TG,i}^{\min}(t), & P_{TG,i}^{\min}(t) > P_{TG,i}(t) \\
P_{TG,i}(t) = \dfrac{\lambda_{TG,i}(t) - b_{TG,i}}{2a_{TG,i}}, & P_{TG,i}^{\min}(t) \le P_{TG,i}(t) \le P_{TG,i}^{\max}(t) \\
P_{TG,i}(t) = P_{TG,i}^{\max}(t), & P_{TG,i}(t) > P_{TG,i}^{\max}(t)
\end{cases}
\tag{32}
$$

When considering the power output climbing constraint, the generator node needs to make the following adjustments

$$
\begin{cases}
P_{TG,i}(t) = \dfrac{\lambda_{TG,i}(t) - b_{TG,i}}{2a_{TG,i}}, & -\Delta P^d_{TG,i} \le P_{TG,i}(t) - P_{TG,i}(t-1) \le \Delta P^u_{TG,i} \\
P_{TG,i}(t) = P_{TG,i}(t), & \begin{aligned} &-\Delta P^d_{TG,i} > P_{TG,i}(t) - P_{TG,i}(t-1) \\ &\text{or } P_{TG,i}(t) - P_{TG,i}(t-1) > \Delta P^u_{TG,i} \end{aligned}
\end{cases}
\tag{33}
$$

When considering the operating constraints and SOC constraints of the energy storage unit, the output power update equation can be expressed as

$$
\begin{cases}
P_{BS,i}(t) = \underline{P}_{BS,i}, & \underline{P}_{BS,i} > P_{BS,i}(t)\, \&\, SOC^{\min}_{BS,r} \le SOC_{BS,r}(t) \le SOC^{\max}_{BS,r} \\
P_{BS,i}(t) = \dfrac{\lambda_{BS,i}(t) - b_{BS,i}}{2a_{BS,i}}, & \begin{aligned} &\underline{P}_{BS,i} \le P_{BS,i}(t) \le \overline{P}_{BS,i} \\ &\&\, SOC^{\min}_{BS,r} \le SOC_{BS,r}(t) \le SOC^{\max}_{BS,r} \end{aligned} \\
P_{BS,i}(t) = \overline{P}_{BS,i}, & \overline{P}_{BS,i} > \overline{P}_{BS,i}\, \&\, SOC^{\min}_{BS,r} \le SOC_{BS,r}(t) \le SOC^{\max}_{BS,r} \\
P_{BS,i}(t) = 0, & SOC^{\min}_{BS,r} > SOC_{BS,r}(t) \cap SOC_{BS,r}(t) > SOC^{\max}_{BS,r}
\end{cases}
\tag{34}
$$

The update equation of the wind generator node with output constraint of wind power is

$$
\begin{cases}
P_{WT,i}(t) = 0, & 0 > P_{WT,i}(t) \\
P_{WT,i}(t) = \dfrac{\lambda_{WT,i}(t) - b_{WT,i}}{2a_{WT,i}}, & 0 \le P_{WT,i}(t) \le P^{st}_{WT,i}(t) \\
P_{WT,i}(t) = P^{st}_{WT,i}(t), & P_{WT,i}(t) > P^{st}_{WT,i}(t)
\end{cases}
\tag{35}
$$

The update equation of the PV node with output constraint of photovoltaic power is

$$
\begin{cases}
P_{SL,i}(t) = 0, & 0 > P_{SL,i}(t) \\
P_{SL,i}(t) = \dfrac{\lambda_{SL,i}(t) - b_{SL,i}}{2a_{SL,i}}, & 0 \le P_{SL,i}(t) \le P^{st}_{SL,i}(t) \\
P_{SL,i}(t) = P^{st}_{SL,i}(t), & P_{SL,i}(t) > P^{st}_{SL,i}(t)
\end{cases}
\tag{36}
$$

When considering the AC/DC connection line constraint, the update equation for the DC-side power deviation is

$$
\begin{cases}
\Delta P_{DC}(t) = P^{\min}_{AC\_DC}, & P^{\min}_{AC\_DC} > P_{AC\_DC}(t) \\
\begin{aligned} \Delta P_{DC}(t) = &\sum_{s \in S_{DCDM}} P_{DCDM,s}(t) - \sum_{k \in S_{SL}} P_{SL,k}(t) \\ &- \sum_{r \in S_{BS}} P_{BS,r}(t) - P_{AC-DC}(t), \end{aligned} & P^{\min}_{AC\_DC} \le P_{AC\_DC}(t) \le P^{\max}_{AC\_DC} \\
\Delta P_{DC}(t) = P^{\max}_{AC\_DC}, & P_{AC\_DC}(t) > P^{\max}_{AC\_DC}
\end{cases}
\tag{37}
$$

When considering the AC/DC connection line constraint, the update equation for the AC-side power deviation is

$$
\begin{cases}
\Delta P_{AC}(t) = P^{\min}_{AC\_DC}, & P^{\min}_{AC\_DC} > P_{AC\_DC}(t) \\
\begin{aligned} \Delta P_{AC}(t) = &\sum_{s \in S_{ACDM}} P_{ACDM,s}(t) - \sum_{i \in S_{TG}} P_{TG,i}(t) \\ &- \sum_{j \in S_{WT}} P_{WT,j}(t) + P_{AC-DC}(t), \end{aligned} & P^{\min}_{AC\_DC} \le P_{AC\_DC}(t) \le P^{\max}_{AC\_DC} \\
\Delta P_{AC}(t) = P^{\max}_{AC\_DC}, & P_{AC\_DC}(t) > P^{\max}_{AC\_DC}
\end{cases}
\tag{38}
$$

By combining Equations (32)–(38) with (27), for the hybrid microgrid, the two-layer consensus algorithm is developed with constraints, which can be used to effectively solve the hierarchical ED problems of the considered hybrid microgrid. It is worth pointing out that, in the hierarchical framework, the generator nodes of the leadership layer and the tracking layer are obtained. For the leadership layer nodes, they can share the feedback data to those adjacent nodes, and for the tracking nodes, they can receive data from the interconnected leader nodes; finally, the microgrid can achieve the consensus state.

For convenience of understanding, the flow chart of the proposed two-layer consensus algorithm is shown in Figure 2.

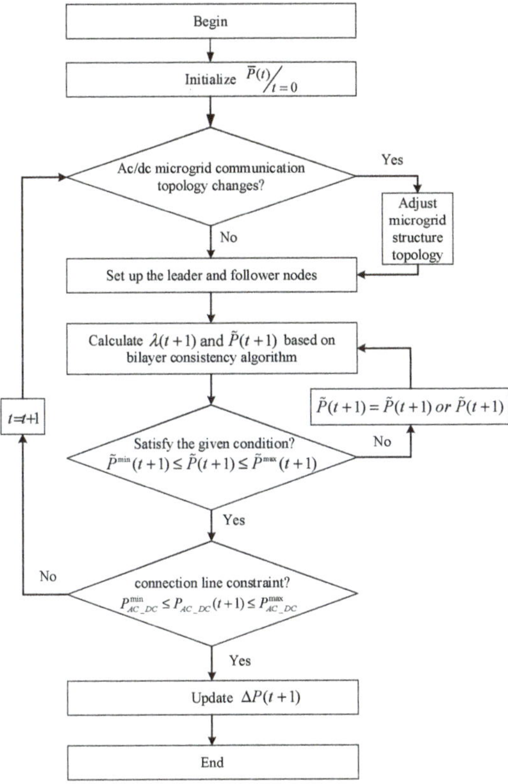

Figure 2. Flow chart of the dynamic economic dispatch (ED) strategy based on the two-layer consensus algorithm.

Remark 1. *It should be pointed out that for the considered hybrid microgrid, a fully distributed hierarchical algorithm based on combining Equations (32)–(38) with (27) is proposed for solving the ED problem, where the necessary constraints for distributed units are also considered, which could be helpful for ED of the considered microgrid. Compared with the existing consensus method-based approaches [39,40], the proposed hierarchical consensus algorithm has a better adaptability. In addition, in some cases where the ED issue has the hierarchical property, the results in References [39,40] would fail in solving this problem.*

Remark 2. *The main difference between this paper and the existing results [39,40,42] is that the hierarchical control thought is utilized in this paper. By the use hierarchical processing, the whole control strategy is divided into two parts, that is, the leadership layer control and the tracking layer control. In our proposed hierarchical consensus strategy, the leadership layer is taken as the upper level, and correspondingly, the tracking layer is taken as the lower level, thus the proposed algorithm could be processed in a hierarchical way, which is in line with the published results. With more leader nodes existing simultaneously, the convergence speed could be faster, and the simulation curves are smoother and more satisfactory.*

5. Case Study

In this paper, without loss of generality, the scheduling time for the considered hybrid microgrid is set as 24 h, and the dispatch period is set as 1 h. In a single period, the load demand is taken as the constant. Before explaining the simulation results, some parameters that were used are illustrated in Table 1.

Table 1. Daily load, wind generator output and photovoltaic (PV) output.

No.	Time Period	Load/kW	PV/kW	Wind Generator/kW
1	00:00–01:00	250	0	83
2	01:00–02:00	200	0	65
3	02:00–03:00	190	0	84
4	03:00–04:00	150	0	62
5	04:00–05:00	160	0	138
6	05:00–06:00	150	0	59
7	06:00–07:00	190	15	143
8	07:00–08:00	250	35	89
9	08:00–09:00	350	95	54
10	09:00–10:00	400	130	84
11	10:00–11:00	350	150	123
12	11:00–12:00	400	180	120
13	12:00–3:00	460	170	104
14	13:00–14:00	410	150	118
15	14:00–15:00	310	120	139
16	15:00–16:00	290	95	55
17	16:00–17:00	300	70	80
18	17:00–18:00	360	50	54
19	18:00–19:00	650	10	69
20	19:00–20:00	720	0	122
21	20:00–21:00	720	0	122
22	21:00–22:00	600	0	137
23	22:00–23:00	500	0	108
24	23:00–24:00	360	0	57

Furthermore, the power generation cost and some operating parameters for the power supply are shown in Table 2 [44].

Table 2. Microgrid system operating parameters.

No.	Node	a_i	b_i	c_i	$[P_i^{\min}, P_i^{\max}]$ (kW)
1	WT	0.0328	7.75	220	[0, 300]
2	TG1	0.0430	7.80	208	[0, 500]
3	BS1	0.0790	7.62	172	[−100, 100]
4	PV	0.0344	7.84	352	[0, 300]
5	TG3	0.0736	7.80	178	[0, 500]
6	BS2	0.0790	7.70	175	[−100, 100]

5.1. Single-Period Simulation Analysis

In the single-period simulation, we verified the proposed algorithm by choosing the data in the time range of 11:00–12:00 in Table 1. The leader nodes are 1, 2, 3, and 5, and the tracking layer nodes are 4 and 6. When considering the constraint conditions shown in Table 2, the proposed distributed two-layer consensus algorithm was tested in the circumstances of isolated AC/DC hybrid microgrid, AC sub-grid, and DC sub-grid, respectively.

5.1.1. Simulation for AC/DC Hybrid Microgrid

For the isolated AC/DC hybrid microgrid, the simulation results are illustrated in Figures 3 and 4. Figure 3 shows the power output of each node of a single period in the isolated AC/DC hybrid microgrid, which finally converges to 136.18, 13.18, 56.54, 129.85, 7.71 and 56.54 kW, and the output of all nodes operate within the corresponding constraint range. Figure 4 shows the power mismatch of a single period in the hybrid microgrid, where we can find that the power supply meets the load demand at the 20th iteration.

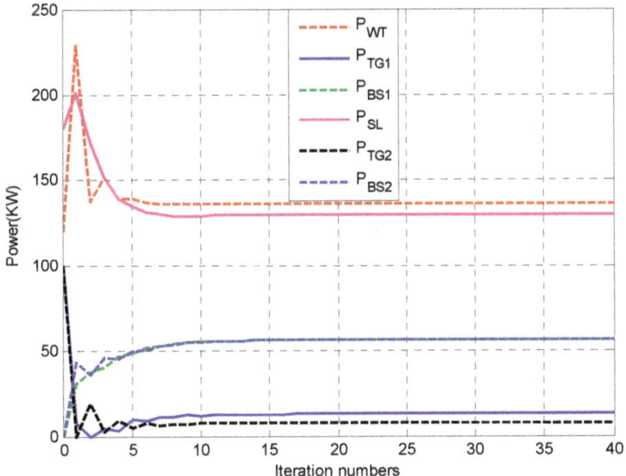

Figure 3. Power output of each node of a single period in the AC/DC hybrid microgrid.

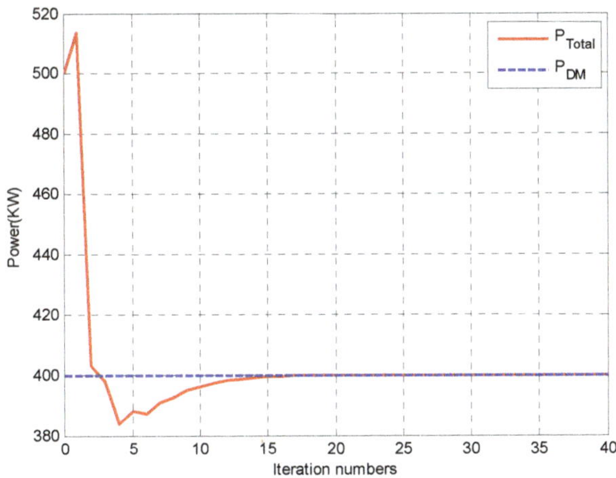

Figure 4. Power mismatch of a single period in the AC/DC hybrid microgrid.

Remark 3. *In the simulation of a single period, the first circumstance we discussed is the AC sub-grid and DC sub-grid, which were investigated as a whole. By calculating the power demand in different sub-grids, the active power balance can be realized by reasonably assigning the power through the tie line. If the active power of the AC sub-grid cannot create self-sufficiency, the missing power will be supplied by the DC sub-grid, and vice versa. The whole active power between the two sub-grids will always keep a dynamic balance, which means the power balance is realized.*

5.1.2. Simulation for AC Sub-Grid

In this case, the wind generator, traditional generator, and energy storage unit are in the AC sub-grid. In the isolated AC/DC hybrid microgrid load from 11:00 to 12:00 shown in Table 1, the load of the AC sub-grid is set as 210 kW. By taking the AC sub-grid as the research object, the simulation results obtained by using the proposed two-layer consensus algorithm are given in Figures 5 and 6. Figure 5 shows the power output of the AC side node of a single period in the AC sub-grid, which finally converges to 138.09, 14.60, and 57.31 kW, respectively. Figure 6 is power mismatch of a single period in the AC sub-grid. We know that the total power output is 210 kW, which can meet the load demand.

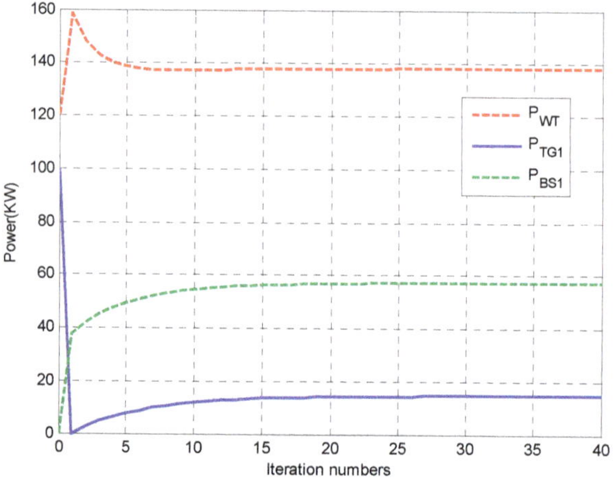

Figure 5. Power output of the AC side node of a single period in the AC sub-grid.

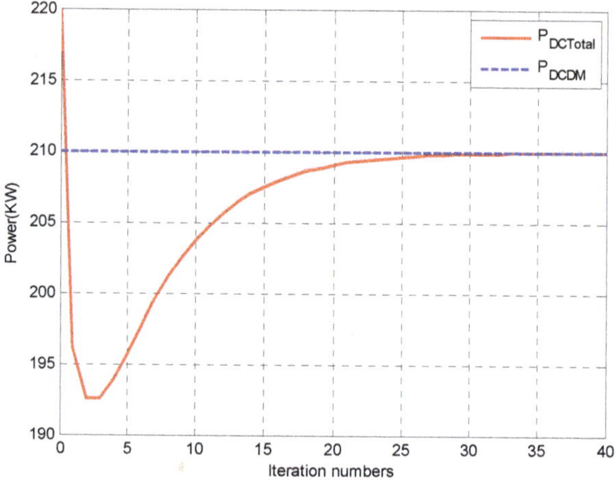

Figure 6. Power mismatch of a single period in the AC sub-grid.

5.1.3. Simulation for DC Sub-Grid

Similarly, three parts of distributed power supply, for instance, photovoltaic power generation, traditional generator, and energy storage unit, are in the DC sub-grid. In the single-period simulation process, the load of the DC sub-grid is taken as 190 kW, and the simulation by using the proposed two-layer consensus algorithm is presented in Figures 7 and 8. Figure 7 is the power output of the DC side node of a single period in the DC sub-grid, which finally converges to 127.70, 6.70, and 55.60 kW respectively, and the sum is 190 kW. Figure 8 is a power mismatch of a single period in the DC sub-grid.

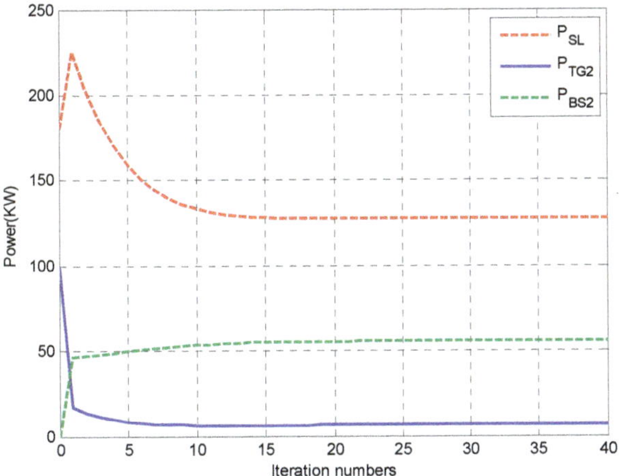

Figure 7. Power output of the DC side node of a single period in the DC sub-grid.

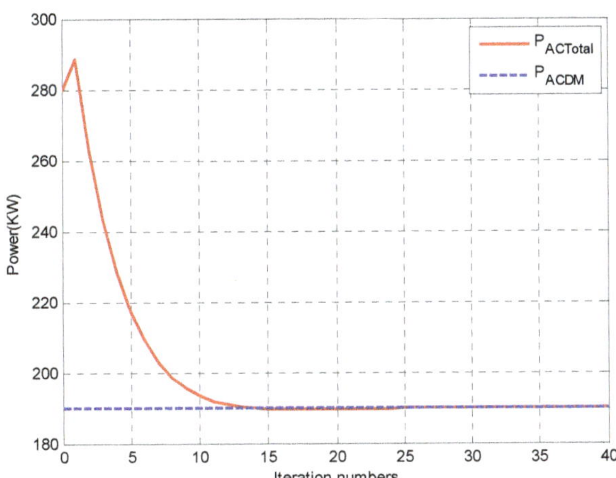

Figure 8. Power mismatch of a single period in the DC sub-grid.

Remark 4. *The second and third circumstances of a single period investigate the power mismatch of the AC sub-grid and the DC sub-grid, respectively. In the AC sub-grid, the power first needs to realize self-sufficiency. That is to say, the total power output of P_{WT} and P_{TG1} must meet the total loads in this section. The redundant power can be reserved in the energy storage units, or the energy storage units release power to supply the power demand of the AC sub-grid. It is a similar criteria to the DC sub-grid.*

5.2. Multi-Period Simulation Analysis

For further verifying the robustness of the proposed two-layer consensus algorithm, we further added a time period from 12:00 to 13:00 to illustrate the effectiveness. In the multi-period, the load is changed from 400 to 460 kW, and the AC/DC hybrid network is taken as the research object. Figures 9–11 illustrate the corresponding simulation results, respectively.

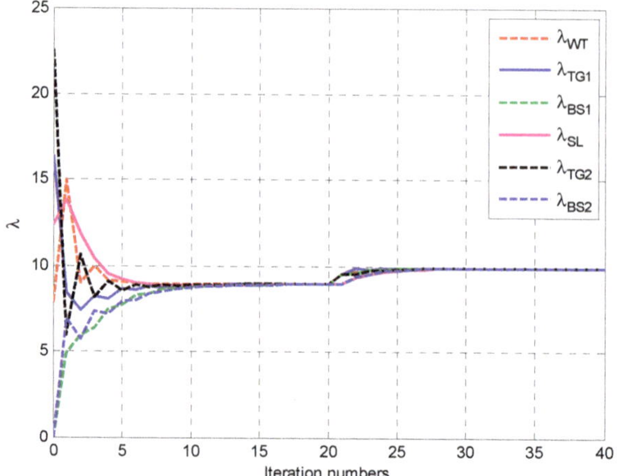

Figure 9. IC update of each node of the multi-period in the AC/DC hybrid microgrid.

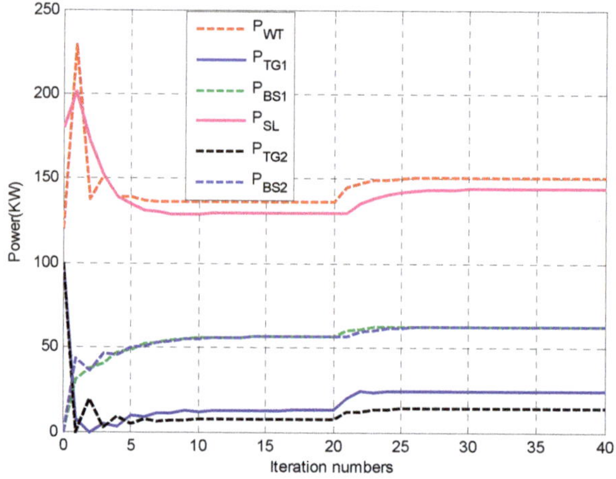

Figure 10. Power output of each node of the multi-period in the AC/DC hybrid microgrid.

The IC update of each node of the multi-period is presented in Figure 9. Before the 20th iteration, the simulation results are the same as those of Figure 3, in the 20th and subsequent iterations, the nodes are abruptly changed due to the load, the IC also changes accordingly, and finally converges to 9.9196. Figure 10 shows the power output of each node of the multi-period. After the load is abrupt, then the output of each node changes accordingly, and finally converges to 151.21, 24.65, 62.78, 144.18, 14.40, and 62.78 kW, respectively. The sum is 460 kW, which just meets the load demand. Figure 11 is the power mismatch of the multi-period in the hybrid microgrid, following which we can find that at each time period, the total power output meets the total load demand.

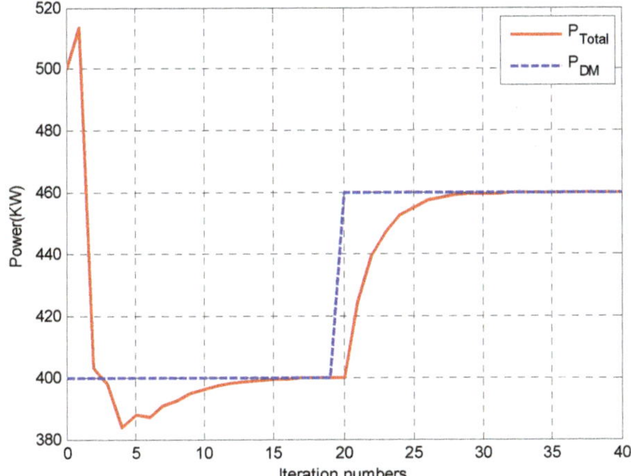

Figure 11. Power mismatch of the multi-period in the AC/DC hybrid microgrid.

5.3. Full Time-Period Simulation Analysis

The proposed two-layer consensus strategy was further verified in a full time period (24 h). By using the data from Tables 1 and 2, the simulation results obtained by the developed two-layer consensus method are provided in Figure 12. From this figure, we know that regardless of the time period, the power output can satisfy the total load demand, so we can conclude that the proposed algorithm has better robustness and adaptability.

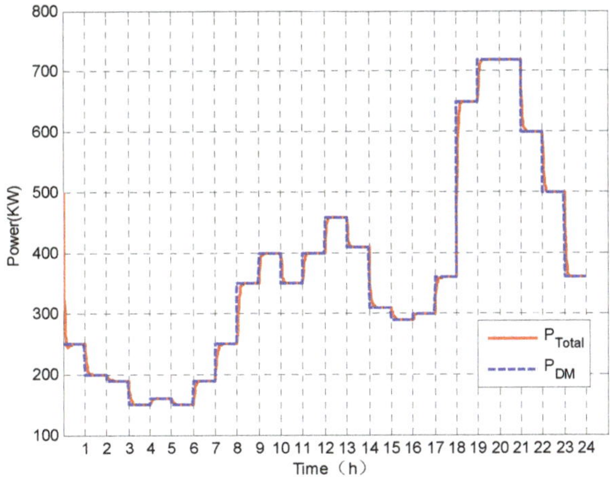

Figure 12. Power mismatch of the full time period (24 h) in the AC/DC hybrid microgrid.

In the same conditions, by using the methods developed in References [30,31], they can all realize the power balance in a short time. However, due to the fact that no nodes were hierarchically controlled in the mentioned results, the interaction times were longer, and the fluctuations were larger, thus we can conclude that the proposed hierarchical consensus algorithm could be much more effective and satisfactory.

Remark 5. *In the circumstances of a multi-period simulation and a full time-period simulation, two or more time periods were considered to verify the proposed algorithm. Although there exists a time span, the power balance in the AC/DC hybrid microgrid can also be realized. It should be noted that at the moment of time varying between the two time spans, the total load power will mutate to another value, and as time goes by, the new power demand can be satisfied by the output power of the generator, PV, and WT. Finally, the total active power reaches the dynamic balance at each time-steady state.*

5.4. Comparison with the Existing Results

In order to verify the robustness and adaptability of the proposed algorithm, some comparison analyses will be given.

5.4.1. Comparison results

In this section, the proposed consensus algorithm will be compared with Reference [42]. Without loss of generality, the parameter values of a and b use the ones in Reference [42] and we chose the time period 12:00–13:00 as an example. From Table 1, we can see that in this period, the load demand is 460 kW. By using the proposed method, the simulation results are shown in Figures 13 and 14.

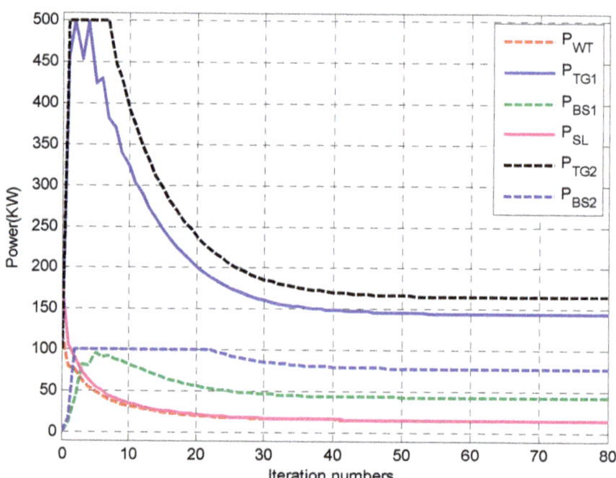

Figure 13. Power output of the proposed method with the same parameters.

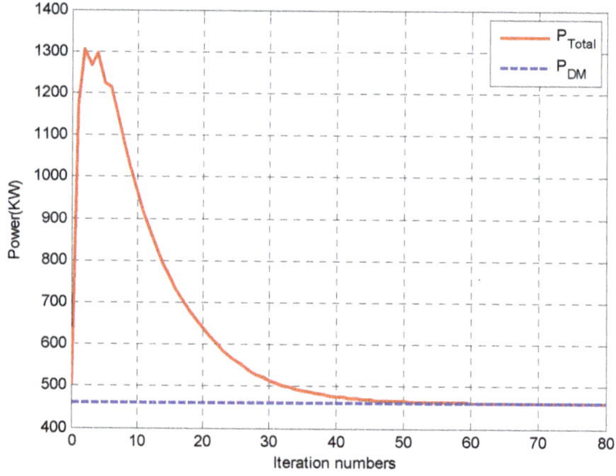

Figure 14. Power mismatch of the proposed method with the same parameters.

It can be seen from Figures 13 and 14 that after about the 60th iteration, the power demand can be satisfied by the output power of the generators, and compared with Reference [42], the proposed method can quickly reach convergence and a smooth transition curve. Besides, each node operates within its constraint condition.

5.4.2. Comparison with Different Parameters

In this section, the proposed method will be tested with the parameters shown in Table 2. The time period we chose is also 12:00–13:00. The simulation results are given by Figures 15 and 16.

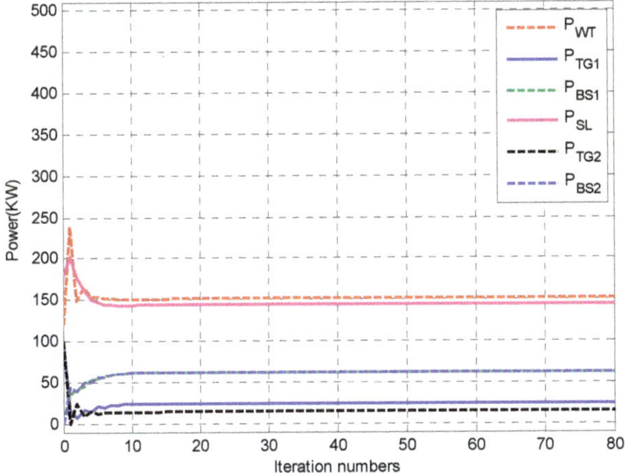

Figure 15. Power output of the proposed method with different parameters.

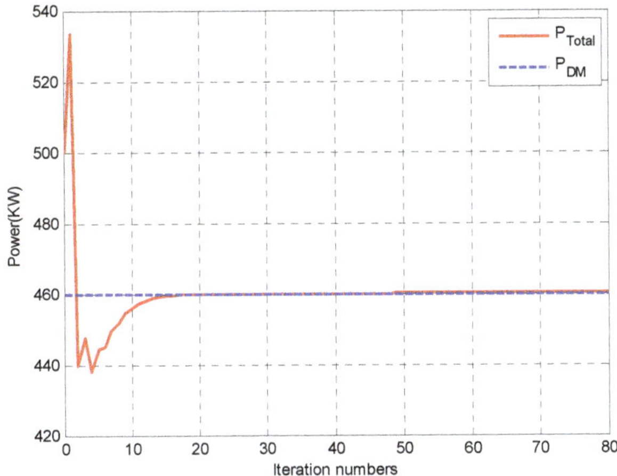

Figure 16. Power output of the proposed method with different parameters.

Compared with Figures 13 and 14, the results in Figures 15 and 16 show better convergence performance, which means that the parameters of a and b can affect the iteration numbers, but cannot change the convergence performance. Thus, the proposed algorithm has a better robustness and adaptation.

6. Conclusions

In this paper, the ED problem of the isolated AC/DC hybrid microgrid was investigated based on the distributed hierarchical consensus method. By using the proposed algorithm, the active power dispatch of the AC/DC microgrid can be realized in a distributed way. Besides, the AC sub-grid and the DC sub-grid can also realize the dynamic power balance in respective sections. If the active power in the AC sub-grid or the DC sub-grid is out of balance, the energy storage units will absorb/release relevant power to ensure the power balance by their charging/discharging process. Finally, some case studies were listed to illustrate the effectiveness of the proposed method, and some comparisons and analyses with the existing results were also provided to verify the robustness and adaptation of the proposed algorithm.

The main work in this paper is realizing the ED problem of the AC/DC hybrid microgrid, which emphatically contains active power dispatch of the AC/DC hybrid microgrid and does not relate to the reactive power dispatch and fault conditions. In future work, the reactive power dispatch problem and the fault conditions of the AC/DC hybrid microgrid will be investigated.

Author Contributions: K.J. developed the concept, conceived the experiments, designed the study, and wrote the original manuscript; K.J. and L.S. performed the experiments and evaluated the data; F.W., L.S., and K.L. reviewed and edited the manuscript. All authors have read and agreed to the published version of the manuscript.

Funding: This research was funded by the National Science Foundation of China (51422701), the Innovation Team of Six Talent Peaks Project of Jiangsu Province (2019-TD-XNY-001), China '111' project of "Renewable Energy and Smart Grid" (B14022).

Conflicts of Interest: The authors declare no conflict of interest.

References

1. Hatziargyriou, N.; Asano, H.; Iravani, R.; Marnay, C. Microgrids. *IEEE Power Energy Mag.* **2007**, *5*, 78–94. [CrossRef]
2. Sao, C.K.; Lehn, W. Control and power management of converter fed microgrids. *IEEE Trans. Power Syst.* **2008**, *23*, 1088–1098. [CrossRef]
3. Akinyele, D.; Belikov, J.; Levron, Y. Challenges of microgrids in remote communities: A steep model application. *Energies* **2018**, *11*, 432. [CrossRef]
4. Mohamed, S.; Shaaban, M.F.; Ismail, M.; Serpedin, E.; Qaraqe, K.A. An efficient planning algorithm for hybrid remote microgrids. *IEEE Trans. Sustain. Energy* **2019**, *10*, 257–267. [CrossRef]
5. Wang, C.; Wu, Z.; Li, P. Research on key technologies of microgrid. *Trans. China Electrotech. Soc.* **2014**, *29*, 1–12.
6. Hooshyar, A.; Iravani, R. Microgird protection. *Proc. IEEE* **2017**, *105*, 1332–1353. [CrossRef]
7. Yazdanian, M.; Mehrizi-Sani, A. Distributed control techniques in microgrids. *IEEE Trans. Smart Grid* **2014**, *5*, 2901–2909. [CrossRef]
8. Song, Y.; Guo, F.; Wen, C. *Distributed Control and Optimization Technologies in Smart Grid Systems*; CRC Press: Boca Raton, FL, USA, 2017.
9. Xu, Y.; Li, Z. Distributed optimal resource management based on the consensus algorithm in a microgrid. *IEEE Trans. Power Electron.* **2015**, *62*, 2584–2592. [CrossRef]
10. Persis, C.D.; Weitenberg, E.R.; Dörfler, F. A power consensus algorithm for DC microgrids. *Automatica* **2018**, *89*, 364–375.
11. Meng, L.; Zhao, X.; Tang, F.; Savaghebi, M.; Dragicevic, T.; Vasquez, J.C.; Guerrero, J.M. Distributed voltage unbalance compensation in islanded microgrids by using a dynamic consensus algorithm. *IEEE Trans. Power Electron.* **2016**, *31*, 827–838. [CrossRef]
12. Schiffer, J.; Seel, T.; Raisch, J.; Sezi, T. Voltage stability and reactive power sharing in inverter-based microgrids with consensus-based distributed voltage control. *IEEE Trans. Control Syst. Technol.* **2016**, *24*, 96–109. [CrossRef]
13. Meng, L.; Dragicevic, T.; Roldán-Pérez, J.; Vasquez, J.C.; Guerrero, J.M. Modeling and sensitivity study of consensus algorithm-based distributed hierarchical control for DC microgrids. *IEEE Trans. Smart Grid* **2016**, *7*, 1504–1515. [CrossRef]

14. Bidram, A.; Davoudi, A. Hierarchical structure of microgrids control system. *IEEE Trans. Smart Grid* **2012**, *3*, 1963–1976. [CrossRef]

15. Shafiee, Q. Robust networked control scheme for distributed secondary control of islanded microgrids. *IEEE Trans. Ind. Electron.* **2014**, *61*, 5363–5374. [CrossRef]

16. Bidram, A.; Davoudi, A.; Lewis, F.L.; Guerrero, J.M. Distributed cooperative secondary control of microgrids using feedback linearization. *IEEE Trans. Power Syst.* **2013**, *28*, 3462–3470. [CrossRef]

17. Zhang, G.; Li, C.; Qi, D.; Xin, H. Distributed estimation and secondary control of autonomous microgrid. *IEEE Trans. Power Syst.* **2017**, *32*, 989–998. [CrossRef]

18. Chen, G.; Feng, E. Distributed secondary control and optimal power sharing in microgrids. *IEEE/CAA J. Autom. Sin.* **2015**, *2*, 304–312.

19. Liu, X.; Wang, P.; Loh, P. A hybrid AC/DC microgrid and its coordination control. *IEEE Trans. Smart Grid* **2011**, *2*, 278–286.

20. Navarro-Rodríguez, Á.; García, P.; Georgious, R.; García, J. Adaptive active power sharing techniques for DC and AC voltage control in a hybrid DC/AC microgrid. *IEEE Trans. Ind. Appl.* **2019**, *55*, 1106–1116. [CrossRef]

21. Eajal, A.; Yazdavar, A.; El-Saadany, E.; Ponnambalam, K. On the loadability and voltage stability of islanded AC-DC hybrid microgrids during contingencies. *IEEE Syst. J.* **2019**, *13*, 4248–4259. [CrossRef]

22. Nejabatkhah, F.; Li, Y. Overview of power management strategies of hybrid AC/DC microgrid. *IEEE Trans. Power Electron.* **2015**, *30*, 7072–70789. [CrossRef]

23. Eghtedarpour, N.; Farjah, E. Power control and management in a hybrid AC/DC microgrid. *IEEE Trans. Smart Grid* **2014**, *5*, 1494–1505. [CrossRef]

24. Xia, Y.; Wei, W.; Yu, M.; Wang, X.; Peng, Y. Power management for a hybrid AC/DC microgrid with multiple subgrids. *IEEE Trans. Power Electron.* **2018**, *33*, 3520–3533. [CrossRef]

25. Zhou, J.; Zhang, H.; Sun, Q.; Ma, D.; Huang, B. Event-based distributed active power sharing control for interconnected AC and DC microgrids. *IEEE Trans. Smart Grid* **2018**, *9*, 6815–6828. [CrossRef]

26. Yoo, H.Y.; Nguyen, T.T.; Kim, H.M. Consensus-based distributed coordination control of hybrid AC/DC microgrids. *IEEE Trans. Sustain. Energy* **2020**, *11*, 629–638. [CrossRef]

27. Melath, G.; Rangarajan, S.; Agarwal, V. A novel control scheme for enhancing the transient performance of an islanded hybrid AC-DC microgrid. *IEEE Trans. Power Electron.* **2019**, *34*, 9644–9654. [CrossRef]

28. Li, X.; Guo, L.; Li, Y.; Guo, Z.; Hong, C.; Zhang, Y.; Wang, C. A unified control for the DC-AC interlinking converters in hybrid AC/DC microgrids. *IEEE Trans. Smart Grid* **2018**, *9*, 6540–6553. [CrossRef]

29. Rousis, A.; Konstantelos, L.; Strbac, G. A planning model for a hybrid AC-DC microgrid using a novel GA/AC OPF algorithm. *IEEE Trans. Power Syst.* **2020**, *35*, 227–237. [CrossRef]

30. Aprlia, E.; Meng, K.; Hosani, M.; Zeineldin, H.; Dong, Z. Unified power flow algorithm for standalone AC/DC hybrid microgrids. *IEEE Trans. Smart Grid* **2019**, *10*, 639–649. [CrossRef]

31. Zhang, Z.; Chow, M. Convergence analysis of the incremental cost consensus algorithm under different communication network topologies in a smart grid. *IEEE Trans. Power Syst.* **2013**, *27*, 1761–1768. [CrossRef]

32. Zhao, C.; He, J.; Cheng, P.; Chen, J. Analysis of consensus-based distributed economic dispatch under stealthy attacks. *IEEE Trans. Ind. Electron.* **2018**, *64*, 5107–5117. [CrossRef]

33. Hamdi, M.; Chaoui, M.; Idoumghar, L.; Kachouri, A. Coordinated consensus for smart grid economic environmental power dispatch with dynamic communication network. *IET Gener. Transm. Distrib.* **2018**, *12*, 2603–2613. [CrossRef]

34. Yang, S.; Tan, S.; Xu, J. Consensus based approach for economic dispatch problem in a smart grid. *IEEE Trans. Power Syst.* **2013**, *28*, 4416–4426. [CrossRef]

35. Tang, Z.; Hill, D.H.; Liu, T. A novel consensus-based economic dispatch for microgrids. *IEEE Trans. Smart Grid* **2018**, *9*, 3920–3922. [CrossRef]

36. Wang, R.; Li, Q.; Zhang, B.; Wang, L. Distributed consensus based algorithm for economic dispatch in a microgrid. *IEEE Trans. Smart Grid* **2019**, *10*, 3630–3640. [CrossRef]

37. Chen, G.; Zhao, Z. Delay effects on consensus-based distributed economic dispatch algorithm in microgrid. *IEEE Trans. Power Syst.* **2018**, *33*, 602–612. [CrossRef]

38. Lv, Z.; Wu, Z.; Dou, X.; Zhou, M.; Hu, W. Distributed economic dispatch scheme for droop-based autonomous DC microgrid. *Energies* **2020**, *13*, 404. [CrossRef]

39. He, H.; Han, B.; Xu, C.; Zhang, L.; Li, G.; Wang, K. Optimal management system of hybrid AC/DC microgrid based on consensus protocols. *Electr. Power Autom. Equip.* **2018**, *38*, 138–146.

40. Lin, P.; Jin, C.; Xiao, J.; Li, X.; Shi, D.; Tang, Y.; Wang, P. A distributed control architecture for global system economic operation in autonomous hybrid AC/DC microgrids. *IEEE Trans. Smart Grid* **2019**, *10*, 2603–2617. [CrossRef]

41. Jiang, K.; Wu, K.; Zong, X. Dynamic economic dispatch of AC/DC microgrid based on finite-step consensus algorithm. In Proceedings of the 2019 IEEE Sustainable Power and Energy Conference (iSPEC), Beijing, China, 21–23 November 2019; pp. 1909–1914.

42. Jiang, K.; Wu, F.; Zong, X.; Shi, L.; Lin, K. Distributed dynamic economic dispatch of an isolated AC/DC hybrid microgrid based on a finite-step consensus algorithm. *Energies* **2019**, *12*, 4637. [CrossRef]

43. Zhang, W.; Xu, Y.; Liu, W.; Zang, C.; Yu, H. Distributed online optimal energy management for smart grids. *IEEE Trans. Ind. Informat.* **2015**, *11*, 717–727. [CrossRef]

44. Wood, A.; Wollenberg, B. *Power Generation, Operation, and Control*; John Wiley & Sons: New York, NY, USA, 1996.

Article

Comparison of the Use of Energy Storages and Energy Curtailment as an Addition to the Allocation of Renewable Energy in the Distribution System in Order to Minimize Development Costs

Mateusz Andrychowicz

Institute of Mechatronics and Information Systems, Lodz University of Technology, 90-924 Lodz, Poland; mateusz.andrychowicz@p.lodz.pl

Received: 12 June 2020; Accepted: 15 July 2020; Published: 21 July 2020

Abstract: This paper presents a comparison of the efficiency of energy storage and energy curtailment as an addition to the allocation of renewable energy in the distribution system in order to minimize development costs using a Mixed Integer-Linear Programming (MILP). Energy sources and energy storages are selected, sized and allocated under operational circumstances such as grid congestions and weather conditions. Loads and power units are modeled by daily consumption and generation profiles respectively, to reflect the intermittent character of renewable generation and consumption of energy. The optimization is carried out for a one-year time horizon using twenty-four representative days. The method is verified on three main simulation scenarios and three sub-scenarios for each of them, allowing for the comparison of the efficiency of each used tool. The main scenarios differ in their share of energy from renewable energy sources (RES) in total consumption. In the sub-scenarios, different tools (RES sizing and allocation, energy storages (ES) sizing and allocation and energy curtailment) are used. The results of this research confirm that energy curtailment is a more efficient additional tool for RES sizing and allocation than energy storages. This method can find practical application for Distribution System Operators in elaborating grid development strategies.

Keywords: MILP optimization; allocation; renewable energy sources; system modeling; energy storages; energy curtailment

---

1. Introduction

Under the increasing installed power of renewable energy sources (RES), the purpose of distribution grids is shifting from passive energy delivery only, to energy delivery and energy production, increasing the role and imposing new duties for Distribution System Operators (DSOs) [1]. One of these duties imposes an obligation on DSOs for the elaboration of development plans, which is a key measure for increasing distribution systems' capacity host for increasing RES penetration. Presently, the location of the renewable connection with the distribution network relates to local weather conditions or mounting capabilities. Therefore, wind turbines are usually located in distant places from household areas, while photovoltaics (PVs) are commonly installed on the top of roofs. As a result, energy generation and consumption are not correlated and power flows are increased in both directions. This is especially visible in the case of residential prosumers equipped with PV systems. In the summer, PV peak generation occurs from 12 p.m. to 2 p.m., and it falls to zero when the sun is setting. This is in opposition to the energy consumption pattern in households where demand peaks in the evening and from 12 p.m. to 2 p.m. are low, because usually, most of the consumers are away from their houses. Consequently, the PV energy is not consumed locally, and instead it is injected to the grid, to be transmitted and consumed, i.e., by commercial and industrial loads, which are often distant from residential areas. This power flow inflicts additional losses, because energy is lost twice: In the evening, due to increased demand, and midday, because of PV generation. This leads to grid infrastructure overdevelopment and stranded cost problems for DSOs and power balancing issues for Transmission System Operators (TSO) and utilities which, even now, face the problem of dynamic load changes (residual load curve variation) resulting in increased re-dispatching costs and reduced power system reliability [1]. These aspects are usually omitted in the research of DG allocation, which usually focuses on power loss minimization or the improvement of voltage stability.

The above-presented problems can be inexpensively solved by the proper allocation of the RES and energy storages (ES) in the distribution system and its operation. If the generation profile of the energy source is adjusted to the demand profile, the majority of the DG generation is self-consumed locally and allows for the avoidance of power losses. An example of such a principle can be seen in an office equipped with a rooftop PV system. Peak demand for commercial type load is correlated with PV generation, due to the working hours and intensity of devices, such as air conditioners.

This paper presents a method for the minimization of distribution system development costs through the optimal allocation of renewable power units, energy storages and the operation of them, including operational features of various types of generation units and consumption of versatile load types. Strong points lie in the complex approach for solving the local problem by:

the modelling of loads and power-generating units by energy consumption/generation profiles reflecting operational conditions;
the modelling of optimization variables allowing for the selection of the type of RES, energy storages (ES), its size and connection point in the grid and its operation.

The paper consists of seven sections, starting with the introduction, followed by the state of the art on methods of RES allocation in power systems. The third chapter presents the method proposed by the author, while the fourth section includes assumptions and fifth section includes simulation scenarios. Simulation results and discussions are presented in the sixth section. The paper ends with conclusions and recommendations for further research and Appendix A which displayed capacity structure for each scenario.

2. State of the Art on the Distribution System Development

This section provides an overview of methods for DG allocation under the following criteria: objective function formulation, variables selection, time horizon, resolution, approach (technical, economic, environmental, etc.), problem formulation and used solving methods.

Power loss minimization is one of the most common objective function formulations implemented for DG allocation problems [2–7]. Equally frequently applicable methods take into account aspects of the quality of supply also, including voltage improvement in an objective function [8–16] or setting constraints on voltage distortion [17,18] while others include the reduction of harmonics distortions as well [19,20]. Power loss reduction and the improvement of voltage are also presented in [21], however these measures are represented by sensitivity factors. Another approach is more energy consumer-oriented maximizing welfare [22] (including minimisation of energy costs and carbon emissions) or minimizing costs of power losses, load not supplied and DG installation O&M costs [23,24]. A similar method—minimizing power losses and energy generation costs—is presented in [25], while in [24,26], air pollutants are minimized as well. Some of the papers present methods using DG allocation for the improvement of supply reliability [27–29]. In [30], the authors propose a restoration matrix that presents the priority of supply from DG. Reference [31] also includes a reduction of the grid upgrade costs. Some papers present a multistep approach, hence, in the first step of the method presented in [12], optimization maximizes the benefits from the node point of view, while in the second step, the optimization maximizes the benefits from the entire considered network. In [32], the first step determines localization only, while the second provides an optimal capacity of each DG. The original approach is presented in [33], where authors maximize system loadability (usage of the network infrastructure). The objective of the optimization in [34] is nodal voltage stabilization. Nevertheless, the most common approach focusses on power loss minimization and voltage improvement, which are optimized in many approaches. This is presented in [35], where a much wider review on the optimization of DG allocation methods is provided.

Novelty of the Paper

The novelty of this research lies in the complex long-term optimization of a distribution system using RES sizing and allocation simultaneously, ES sizing and allocation and energy curtailment. Furthermore, the technical aspects of distribution system operation such as generation and consumption profiles, power flows, power losses or voltage levels are included.

The above-listed objectives can be obtained by the allocation of DG which can be carried out in several ways, including versatile types of variables. The significant majority of optimisation methods focus on the DG sizing and placing only [9,17–19,22,25,36,37]. Some methods also consider the selection of the DG type [3,7,20,22,23,31,38] but none of the mentioned methods combine the allocation of RES and ES or the allocation of RES and energy curtailment.

Optimization can be carried out for different time horizons and with different time resolutions. Most of the papers present methods optimising the DG structure for a single moment, omitting the dimension of time [2,5,8–11,17–19,22,23,25,36,39–41]. This static approach does not allow us to include crucial aspects of daily demand changes or intermittent RES generation, which are included in the presented paper. In papers where intermittent RES generation is included—[4] and [7]—there is a lack of long-time planning.

In papers [3,7,20,42,43], which present methods for long-time planning, no distribution system costs were minimized.

Most of the research already performed focuses on technical aspects, such as power line constraints [16] or nodal voltage limits, and only a few of them include economic aspects such as investment and O&M costs [17,23] or total cost of energy [19,22]. Only one of the papers reviewed includes different types of loads connected to considered systems [9]. This is another novel aspect of the paper that combines technical and economic aspects of distribution system operating, including different types of loads and generation units also.

Research also exists which combines the allocation of renewable energy sources and energy storages in the distribution system [44–46]. Other papers describe different approaches to energy curtailment: Turning off generation sources [47], limiting generation to a constant level [48], generation constraints adapted to technical constraints [49] or proportional generation reduction for the whole analyzed period [50]. None of the mentioned papers, however, combine the allocation of RES and ES or the allocation of RES and energy curtailment. Furthermore, none of the papers compare which distribution development strategy is more efficient.

3. Problem Formulation

The objective of the method presented is to provide the structure of the RES and ES allocation in the distribution system along with its operation, ensuring minimal costs of the distribution system development without violating the technical standard of the grid operation. Three types of RES (wind turbine (WT), photovoltaic (PV) and biogas (BG)) can be connected in each node with rated power depending on the value of typical units. RES units are represented by representative generation profiles, reflecting typical operating conditions. Due to the grid, overloads can be identified and eliminated by proper RES allocation and curtailment and energy storing. The optimization is performed, preserving technical constraints. Technical constraints refer to the network transmission capacity, nodal voltage standards and power exchange with the transmission system. Power flow in each line is calculated for each time step. The generation for RES is calculated as a product of the installed capacity and generation profile and load consumption is calculated similarly. Power exchange with the transmission system depends on the power of each transformer which connects these two systems.

The optimization model, which is described in the next section, consists of general mathematical formulations for mixed-integer linear problems which can be used by other optimization software.

3.1. Optimization Model

The minimization of the total costs of the distribution system development in relation to one year (installation and operating RES and energy storages) is the objective function. Total costs consist of three elements:

Fixed costs (CAPEX and OPEXfix) of new RES,
Fixed costs (CAPEX and OPEXfix) of new energy storages,
Operating costs (OPEXvar).

CAPEX there are investment costs connected with equipment and construction works costs. OPEXfix there are costs intended for taxes, maintenances and salaries. OPEXvar are the costs of fuels.

$$\min\left\{\sum_{n\in N}\left(\sum_{d\in D}\left(\sum_{r\in R}\left(p_{n,d,r}*P_{d,r}*FC_{d,r}^{RES}\right)\right)+p_n^{ES}*C^{ES}*FC^{ES}+\sum_{d\in D}\left(E_{n,d}*VC_{d,r}^{RES}\right)\right)\right\} \tag{1}$$

subject to

$$\underset{n\in N}{\forall}\quad\underset{d\in D}{\forall}\quad\underset{r\in R}{\forall}\quad p_{n,d,r}\in N \tag{2}$$

The DG generation in each node for each type of RES ($E_{n,d}$) is calculated as a sum of energy generation of each DG type in one node for the entire year.

$$\underset{n\in N}{\forall}\quad\underset{d\in D}{\forall}\quad E_{n,d}=\sum_{t\in T}E_{n,d,t}^{GEN\_RES} \tag{3}$$

The DG generation in each time period ($E_{n,d,t}^{GEN\text{-}RES}$) is a function of the optimization variable $p_{n,d,r}$ and is calculated a product of the RES rated power ($P_{d,r}$), current power utilization level (according to the generation profile $P_{d,r}^{avail}$), optimization variable $p_{n,d,r}$ and minus energy curtailment ($e_{n,d,t}^{curt}$). The $p_{n,d,r}$ is a three-dimension variable, where dimension n refers to the number of the node, dimension d refers to the RES technology type (i.e., PV) and dimension r refers to the rated power (2). Energy curtailment ($e_{n,d,t}^{curt}$) is a variable which curtails the generation from the reference generation profiles for each node (n), RES type (d) and each time period (t).

$$\underset{n\in N}{\forall}\quad\underset{d\in D}{\forall}\quad\underset{t\in T}{\forall}\quad E_{n,d,t}^{GEN\_RES}=\sum_{r\in R}\left(\left(P_{d,r}\cdot p_{n,d,r}\right)\cdot P_{d,t}^{avail}\right)-e_{n,d,t}^{curt} \tag{4}$$

Another important parameter is the generation in the entire node ($E_{n,t}^{GEN\_node}$) which depends on the sum of DG generation for each RES type in single node and energy exchange with energy storages that are installed in the node.

$$\underset{n\in N}{\forall}\quad\underset{t\in T}{\forall}\quad E_{n,t}^{GEN\_node}=\left(\sum_{d\in D}\left(E_{n,d,t}^{GEN\_RES}\right)-e_{n,t}^{toES}+e_{n,t}^{fromES}\right) \tag{5}$$

The typical RES sizes of each renewable technology are predefined and are included in the two-dimension matrix $P_{d,r}$ (4). As a result of multiplication of variable matrix $p_{n,d,r}$ and parameters matrix $P_{d,r}$ the installed power of all RES in every node is calculated.

$$
\underset{d \in D}{\forall} \quad \underset{r \in R}{\forall} \quad P_{d,r} = \overset{dimmension\ D}{\left[\begin{array}{ccc} P_{1,d_1} & P_{1,d_2} & P_{1,d_3} \\ P_{2,d_1} & P_{2,d_2} & P_{2,d_3} \\ \vdots & \vdots & \vdots \\ P_{r,d_1} & P_{r,d_2} & P_{r,d_3} \end{array} \right]} \Bigg\} \ dimmension\ R
\tag{6}
$$

The optimization model in general is formulated as for an integer programming, however, it includes some non-linear elements such as power losses or voltage drops. In order to improve computation time and reformulate the original problem to the Mixed Integer-Linear Programming (MILP), non-linear components are linearized. The quadratic power losses are linearized by the spline of five linear functions—Figure 1. Each linear function was created for different values of power flow in relation to power line capacity. The first function was created for power flow in rage 0–100% of line capacity. The rest of the functions were designed for power flow in range: $f2$: 20%–100%, $f3$: 40%–100%, $f4$: 60%–100% and $f5$: 80%–100% of line capacity. The final power losses were created as a sum of all mentioned linear functions.

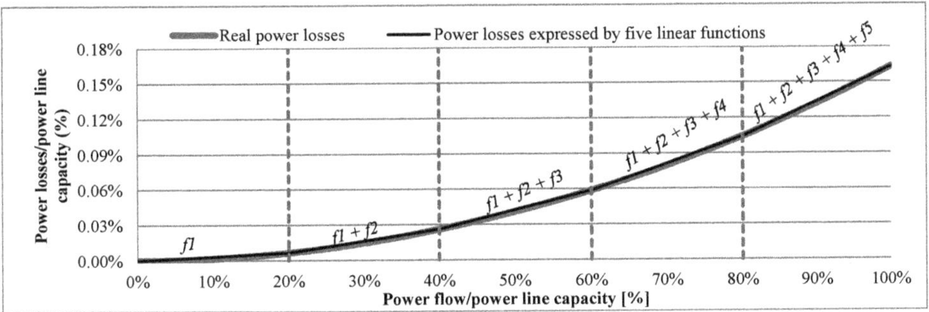

Figure 1. Power losses linearization—spline function consisting of five linear functions.

Each linearising function refers to power line capacity, this is applicable for every power line of the grid model (7)–(11).

$$
\underset{t \in T}{\forall} \quad \underset{n,w \in N}{\forall} \quad f1_{n,w,t} = 0.2 * \frac{Line.const_{n,w} * R_{n,w}}{U_{n,t}{}^2} * PF_{n,w,t}
\tag{7}
$$

$$
\underset{t \in T}{\forall} \quad \underset{n,w \in N}{\forall} \quad f2_{n,w,t} = 0.4 * \frac{Line.const_{n,w} * R_{n,w}}{U_{n,t}{}^2} * PF_{n,w,t} - 0.08 * \frac{Line.const_{n,w}^2 * R_{n,w}}{U_{n,t}{}^2}
\tag{8}
$$

$$
\underset{t \in T}{\forall} \quad \underset{n,w \in N}{\forall} \quad f3_{n,w,t} = 0.4 * \frac{Line.const_{n,w} * R_{n,w}}{U_{n,t}{}^2} * PF_{w,k,t} - 0.16 * \frac{Line.const_{n,w}^2 * R_{n,w}}{U_{n,t}{}^2}
\tag{9}
$$

$$
\underset{t \in T}{\forall} \quad \underset{n,w \in N}{\forall} \quad f4_{n,w,t} = 0.4 * \frac{Line.const_{n,w} * R_{n,w}}{U_{n,t}{}^2} * PF_{w,k,t} - 0.24 * \frac{Line.const_{n,w}^2 * R_{n,w}}{U_{n,t}{}^2}
\tag{10}
$$

$$
\underset{t \in T}{\forall} \quad \underset{n,w \in N}{\forall} \quad f5_{n,w,t} = 0.4 * \frac{Line.const_{n,w} * R_{n,w}}{U_{n,t}{}^2} * PF_{w,k,t} - 0.32 * \frac{Line.const_{n,w}^2 * R_{n,w}}{U_{n,t}{}^2}
\tag{11}
$$

Consequently, the energy lost on the power lines in each time step is as in (12),

$$\underset{t \in T}{\forall} \quad \underset{n,w \in N}{\forall} \quad P^{losses}_{n,w,t} = f1_{n,w,t} + f2_{n,w,t} + f3_{n,w,t} + f4_{n,w,t} + f5_{n,w,t} \tag{12}$$

while the total power losses in the entire time horizon are as (13).

$$\underset{t \in T}{\forall} \quad P^{losses}_{t} = \sum_{n,w \in N} \left(P^{losses}_{n,w,t} \right) \tag{13}$$

The same assumption of constant nodal voltages and omission of reactive power and reactance allows for the linearization of voltage drops (14).

$$\underset{t \in T}{\forall} \quad \underset{n,w \in N}{\forall} \quad \Delta U_{n,w,t} = \frac{PF_{n,w,t} * R_{n,w}}{U_{n,t}} \tag{14}$$

Power flow in a power line, as well as nodal voltage, depends on power balance in each node depending on the grid structure and temporary power generation and demand. While the generation depends on weather conditions and generation structure, the demand for power in each time period depends on the load size and its type (referring to user behaviors).

$$\underset{n \in N}{\forall} \quad \underset{t \in T}{\forall} \quad \sum_{n \in N}(PF_{n,w,t}) = E^{GEN\_node}_{n,t} - \sum_{l \in L}(Load_{n,l} * P^{DEM}_{l,t}) \tag{15}$$

3.2. Constraints

Two types of constraints are included: technical and economic. Technical constraints include power balances in grid nodes (16), power lines capacity (17) and nodal voltage limitations (18).

$$\underset{t \in T}{\forall} \quad \sum_{n \in N} \left\{ \sum_{l \in L}(Load_{n,l} * P^{DEM}_{l,t}) \right\} - P^{res}_{t} + TS^{exchange}_{t} + PF_{n,w,t} = 0 \tag{16}$$

$$\underset{t \in T}{\forall} \quad \underset{n,w \in N}{\forall} \quad -Line.const_{n,w} \leq PF_{n,w,t} \leq Line.const_{n,w} \tag{17}$$

$$\underset{t \in T}{\forall} \quad \underset{n \in N}{\forall} \quad 0.9 * Unominal \leq \underline{U}_{n,t} \leq 1.1 * Unominal \tag{18}$$

In the model, there are also assumed constraints related to the operation of energy storages like: level of charge, which depends on energy exchange between energy storage and node and efficiency of this exchange.

$$n \in N, \forall t \in T \quad CL_{n,t} = CL_{n,t-1} + \eta^{ES} * P^{doES}_{n,t} - \left(\frac{1}{\eta^{ES}} \right) * P^{zES}_{n,t} \tag{19}$$

The model assumes that annual energy production minus total losses in lines must be equal or greater than the assumed share (k) in the annual demand in the distribution system (20).

$$\sum_{n \in N} \left\{ \sum_{d \in D} E_{n,d} - \sum_{t \in T} P^{losses}_{t} \right\} \geq k * \sum_{n \in N} \left\{ \sum_{l \in L} \left(\sum_{t \in T} Load_{n,l} * P^{DEM}_{l,t} \right) \right\} \tag{20}$$

4. Assumptions

Since the model includes natural (non-negative integer) and real variables and all functions are linear, the optimization problem is modeled as a Mixed Integer-Linear Programming (MILP).

The natural variable $p_{n,d,r}$ indicates the type of RES selected from the set of three technologies D: PV systems, wind turbines and biogas thermal units (21).

$$D = \{d_1, d_2, d_3\} = \{pv, wt, bg\} \tag{21}$$

The rated power of each type of the RES is retrieved from the impact assessment to polish regulation on renewable energy sources from 2015 [51], displayed in Table 1. In the case of PV, the author realized that due to modular structure, PV systems are fully configurable, although they are assumed as three predefined values, to reduce the calculation effort. Nevertheless, the method allows for the application of any set of parameters.

Table 1. Rated power of the renewable energy sources (RES).

Type Series	PV	WT	BG
1	10 kW	10 kW	200 kW
2	100 kW	100 kW	500 kW
3	1000 kW	500 kW	2000 kW

There was one assumed type of energy storage with a capacity of 10 kWh, power of 5 kW and efficiency of energy exchange with the grid on the level of 90%.

Generation profiles are divided into three parts of the year: summer, winter and spring/autumn together. For each part of the year, there were created two profiles for WT and PV (high and low generation) based real data retrieved from the website of German and PolishTSOs. Generation profile for BG is assumed as constant for the whole year.

Three types of loads are included: residential, commercial and industrial. Each type of load is also characterized by a different energy consumption profile. Consumption profiles are also divided into three parts of the year and for each part, two profiles were created for working and non-working days. Load location is predefined (Figure 2).

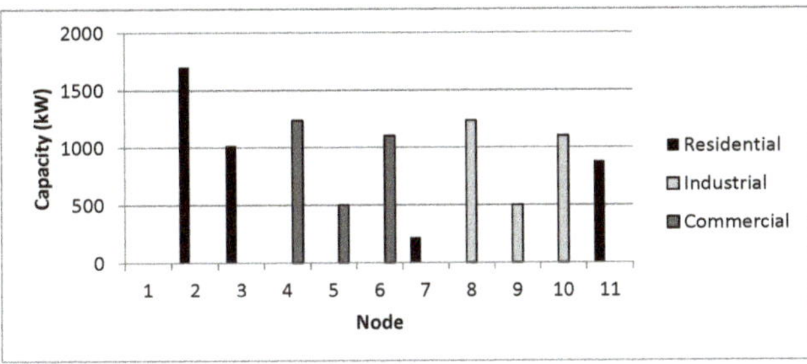

Figure 2. Structure of loads connected to the benchmark network model. Own study.

Simulations were carried out for twenty-four representative days which were created as a total combination of generation profiles and consumption profiles, and according to real data, there was an analysis of how many times each combination occurred in the year, and based off of this, the whole year is modeled.

The modified MV benchmark network model consisting of eleven nodes is used for the simulation—Figure 3. Lines are characterized as connections between nodes, and for each of them, capacity and resistance are assumed. There is one exception, which is between nodes one and two and it is a transformer which connects the transmission (node 1) and the distribution system (node 2). The resistance for the transformer is assumed as 0 ohms, because this part of the system is not the aim of this research.

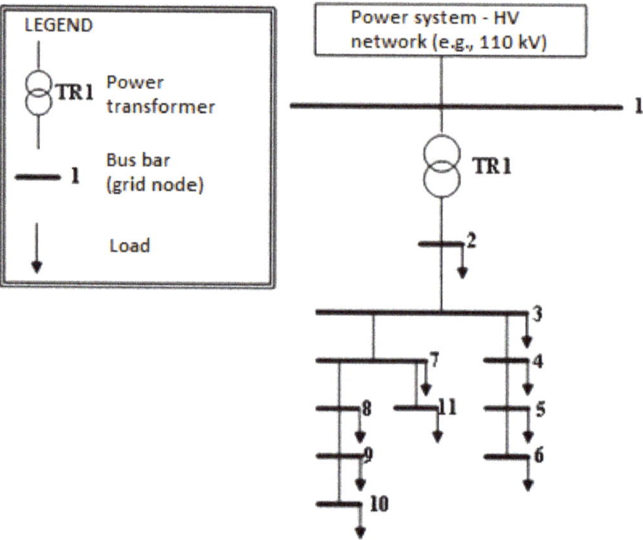

Figure 3. Benchmark model of medium voltage distribution network.

Power lines capacity and resistance are taken from the real data—Table 2.

Table 2. Power lines parameters.

Line	Connection between Nodes	Line Capacity (MW)	Line Resistance (Ω)
1	1–2	40	0 *
2	2–3	7.3	1.21
3	3–4	5.9	0.77
4	3–7	5.9	0.77
5	4–5	5.9	2.89
6	5–6	5.9	1.77
7	7–8	5.9	0.36
8	7–11	5.9	1.77
9	8–9	5.9	0.29
10	9–10	5.9	0.19

* resistance of a power transformer is negligible and assumed as equal to 0.

One of the factors affecting the power balance of the distribution system is power exchange with the transmission system. In this research, it is assumed that power flow is allowed only in one direction—from the transmission to the distribution system. This means that in the period when energy consumption is higher than generation, the difference between them is covered by power supplied by external generation units from the transmission system. However, power cannot flow from the distribution to transmission system. This causes the sum of generation to be lower than the sum of demand in the whole system. This assumption is made arbitrarily and can be removed without affecting the method principle. This will result only with the higher local capacity host for DG.

CAPEX and OPEX (Table 3) depend on the type and the size of a power unit and are assumed on impact assessment to polish the regulation of RES [51].

Table 3. Costs of RES installation.

Type of Cost	Type of RES								
	PV 10	PV 100	PV 1000	WT 10	WT 100	WT 500	BG 200	BG 500	BG 1000
CAPEX [thous. €/MW_year]	84	84	81	110	110	99	214	198	188
OPEXfix [thous. €/MW_year]	27	27	24	36	36	45	144	183	181
OPEXvar [€/MWh]	0	0	0	0	0	0	91	78	70

5. Simulation

5.1. Scenarios

The method is examined in three main simulation scenarios which differ in the share of energy from RES in total demand in the analyzed distribution system. In the first scenario, the share of energy from RES is 30%, in the second 40%, and in the last scenario 50%. Furthermore, each major scenario has three sub-scenarios that were created to display the differences between approaches to the development of distribution systems. In Sub-scenario A, only the sizing and allocation of RES were used, in Sub-scenario B, the sizing and allocation of RES and ES were used, and in the last sub-scenario, C, the sizing and allocation of RES and energy curtailment were used.

Main scenarios:

30% share of energy from RES in total demand,
40% share of energy from RES in total demand,
50% share of energy from RES in total demand.

Sub-scenarios:

sizing and allocation of RES,
sizing and allocation of RES and ES,
sizing and allocation of RES and EC.

5.2. Revision of the Strucure

The minimization of total costs of the distribution system development in relation to one year (installation and operating RES and energy storages) is the objective function. Simulations are performed using a network consisting of eleven nodes, where the node number one represents the transmission system and the rest of the nodes represent the distribution system. In nodes 2–11, generation units and ES can be connected. The connection between nodes is represented by resistance and line capacities. The generation from RES is calculated as a product of the RES rated power, current power utilization level (according to the generation profile), number of units and minus energy curtailment. It is also possible that energy can be stored in an energy storage, which is connected to the nodes.

6. Results

This section provides the simulation results presented in two parts that compare approaches to the development of distribution systems. The first part displays capacity structure for each scenario and the second one displays the overall cost of system development. In each part, the results consist of three charts that present the results for a different share of energy from RES in total demand. All of the results are obtained using FICO®Xpress v. 8.6. optimization software and all graphs are created using Microsoft Excel 365.

Based on Figures 4–6, it can be seen that additional tools to RES sizing and allocation, like energy curtailment (EC) allocation or energy curtailment, allows a user to install more wind turbines, which are the cheapest renewable energy sources, without violating grid technical standards. Energy storages shift energy from one moment to another, which prevents the over-generation from RES. Energy curtailment, on the other hand, fits the generation profile to demand.

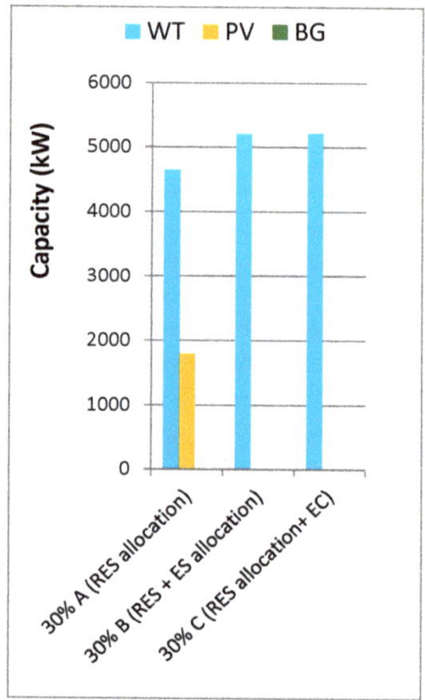

Figure 4. Structure of capacity for main scenarios—1.

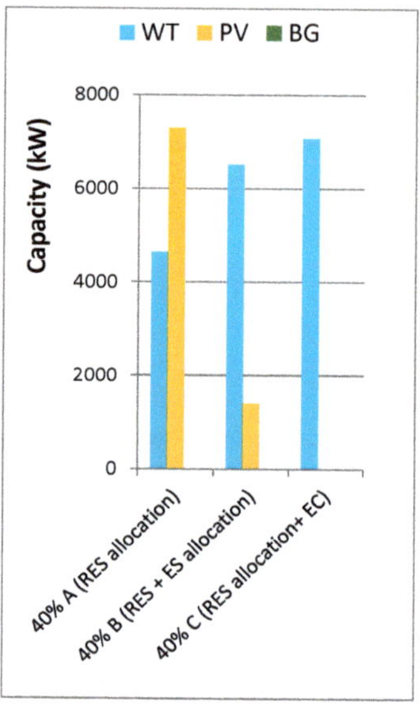

Figure 5. Structure of capacity for main scenarios—2.

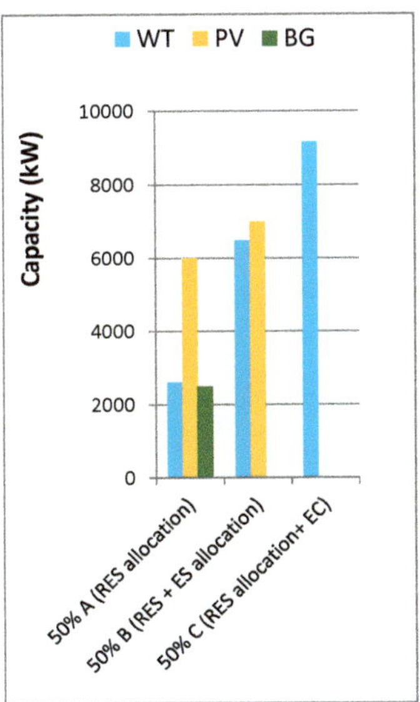

Figure 6. Structure of capacity for main scenarios—3.

Based on Figures 7–9 (all costs are compared to the highest one which is for 50% share for RES with RES sizing and allocation only), it can be observed that the combination of RES sizing and allocation and another additional tool reduces the overall cost of the distribution system development (at least 12% for 30% share). It can also be seen that, together with an increasing share of energy from RES in total demand, the energy curtailment became more efficient than energy storages.

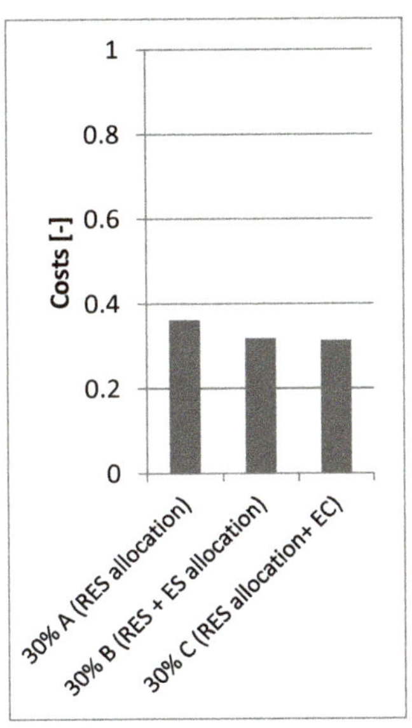

Figure 7. Costs for main scenarios—1.

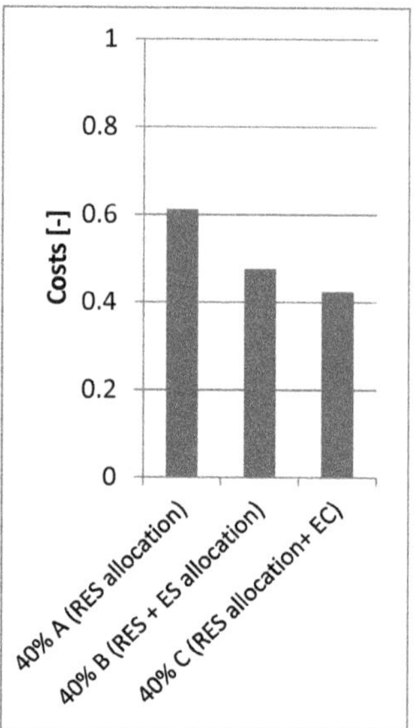

Figure 8. Costs for main scenarios—2.

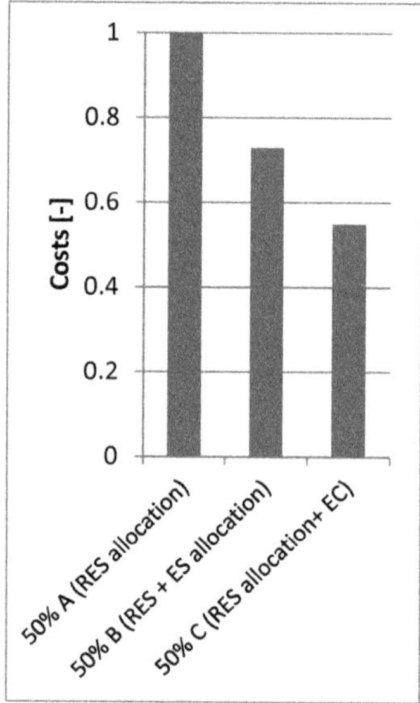

Figure 9. Costs for main scenarios—3.

In Table 4, computation time and the number of variables are displayed for each of the analyzed scenarios.

Table 4. Computation time and number of variables.

Scenario	Computation Time [s]	Number of Variables
30% (RES allocation)	249	17,532,417
30% (RES + ES allocation)	641	17,594,652
30% (RES allocation + EC)	81	17,672,385
40% (RES allocation)	88	17,532,417
40% (RES + ES allocation)	607	17,594,652
40% (RES allocation + EC)	36	17,734,620
50% (RES allocation)	109	17,532,417
50% (RES + ES allocation)	173	17,594,652
50% (RES allocation + EC)	156	17,672,385

Results Discussion

The allocation and sizing of RES allows for the better use of existing infrastructure. This operation consists of choosing the type of renewable energy sources, the number of sources from a given power series and the node in which the units are to be built. Taking into account the generation profiles of individual RES, loads profiles and capacity of existing lines, it is possible to locate such generating units that do not violate the technical parameters of the network, and allow the achievement of the set level of their annual generation at the lowest possible cost total.

The allocation and sizing of energy storage allows for their distribution in the network and the selection of capacity, which will allow them to support RES at times when their generation exceeds the total demand in the system. Energy storage allows for energy storage at times when the generation from generating units exceeds the demand (usually, these are the night demand valleys) and putting this energy into the network at the peak of demand.

Energy curtailment allows a user to reduce energy production at times when their generation exceeds the total demand, which makes it possible to adjust the generation to the demand in these periods.

7. Conclusions

The optimization method for RES, ES allocation and energy curtailment under the criterion of minimization costs of distribution system development is presented. The novelty lies in comparing energy storages (ES) and energy curtailment (EC) as additional tools to RES sizing and allocation.

This combination of mentioned tools allows for the achievement of a desirable target (reducing costs of distribution system development). Due to the modelling of power units with energy generation or consumption profiles, it is possible to consider specific operational cases and select the best structure and location of renewables.

The results also show that energy curtailment is more efficient than energy storages with the growth of the share of energy from RES in total demand. This is due to the high prices of the ES and its limitation connected to the efficiency and level of charge. It shows that energy curtailment can be an efficient tool to control the operation of RES.

Funding: This research received no external funding.

Acknowledgments: Author would like to express the gratitude to the FICO® corporation and Microsoft corporation for programming support and provision of academic licenses for Xpress Optimization Suite and Office 365 to Lodz University of Technology.

Conflicts of Interest: Author declare no conflict of interest.

Abbreviations

Sets

$n, w \in N$	sets of indices n, w representing number of nodes in distribution network
$d \in D$	set of indices d representing type of renewable energy sources (RES) technology—$D = [d_1, \ldots d_3]$, where d_1 is the first possible technology and d_3 is the last one
$l \in L$	set of indices l representing number of possible load type. $L = [l_1, \ldots l_3]$, where l_1 is the first possible type and l_3 is the last one
$r \in R$	set of indices r representing type of rated power for each type of possible RES technology

Coefficients

$E_{n,d}$	annual energy production from each type of RES r in each node n
$E_{n,d,t}^{GEN\_RES}$	total production from RES type d in node n in time t
$E_{n,t}^{GEN\_node}$	total production from RES in node n in time t
p_t^{losses}	total power losses in distribution system in time t
$p_{n,w,t}^{losses}$	power losses in a power line between nodes w and k in time t
$P_{d,r}$	rated power of RES technology d for the power unit series type r
$p_{d,t}^{avail}$	generation profile for RES technology d in time t
$p_{n,w,t}^{active}$	active power flow between nodes n and w in time t
$Load_{n,l}$	nominal power of load type l in node n
$p_{l,t}^{DEM}$	ergy consumption profile (discretized with hourly resolution) of load type l in time t
$PF_{n,w,t}$	linearized power flow between nodes n and w in time t
$TS_t^{exchange}$	energy exchange between distribution and transmission system in time t
$Line.const_{n,w}$	capacity of power line between nodes n and w
$U_{nominal}$	nominal voltage for distribution system (in this paper assumed as a 30 kV)
$R_{n,w}$	resistance of power line between nodes n and w
$U_{n,t}$	active voltage in node n in time t
$\Delta U_{n,w,t}$	linearized value of voltage drop between nodes n and w in time t
$FC_{d,r}^{RES}$	fixed cost of each type and rated power of renewable energy sources
FC^{ES}	fixed cost of energy storages
$VC_{d,r}^{RES}$	variable cost of each type d and rated power r of renewable energy sources
C^{ES}	nominal capacity of single energy storage
$CL_{n,t}$	level of charge of energy storages in node n in time t
η^{ES}	efficiency of energy exchange between nodes and energy storages

Decision variable

$p_{n,d,r}$	number of units in node n for RES type d and rated power r
p_n^{ES}	number of units in node n for energy storages
$e_{n,d,t}^{curt}$	energy curtailment of RES type d in node n in time t
$e_{n,t}^{toES}$	energy which flow from grid to energy storage in node n in time t
$e_{n,t}^{fromES}$	energy which flow from energy storage to grid in node n in time t

Acronyms

WT	wind turbine
PV	photovoltaic installation
BG	biogas power plant
RES	renewable energy sources
ES	energy storages
EC	energy curtailment

Appendix A

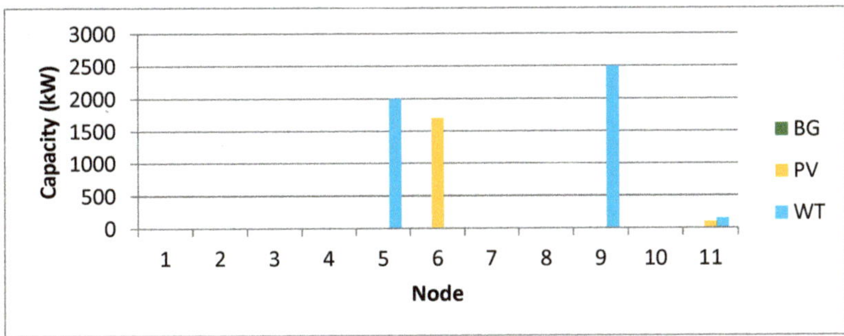

Figure A1. Capacity structure for 30% share of energy from RES in total demand with sizing and allocation of RES only.

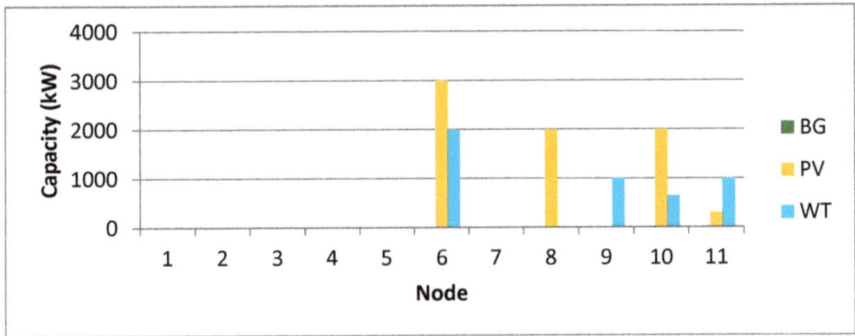

Figure A2. Capacity structure for 40% share of energy from RES in total demand with sizing and allocation of RES only.

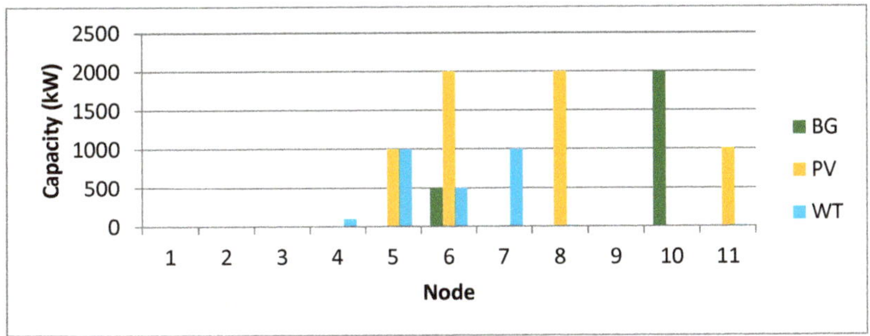

Figure A3. Capacity structure for 50% share of energy from RES in total demand with sizing and allocation of RES only.

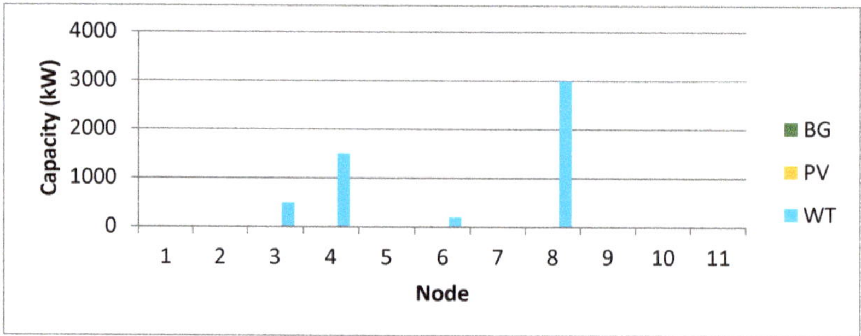

Figure A4. Capacity structure for 30% share of energy from RES in total demand with sizing and allocation of RES and energy storages (ES).

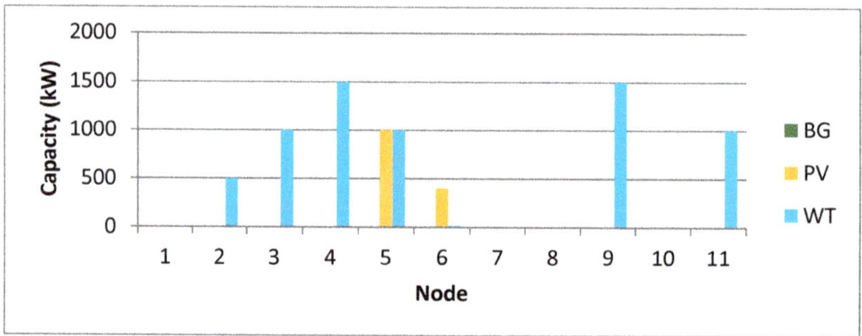

Figure A5. Capacity structure for 40% share of energy from RES in total demand with sizing and allocation of RES and ES.

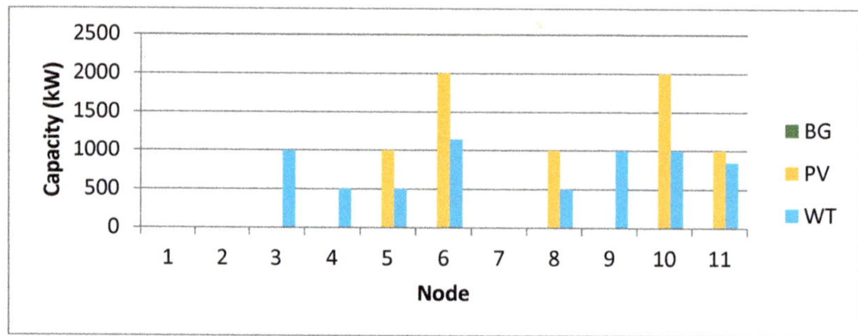

Figure A6. Capacity structure for 50% share of energy from RES in total demand with sizing and allocation of RES and ES.

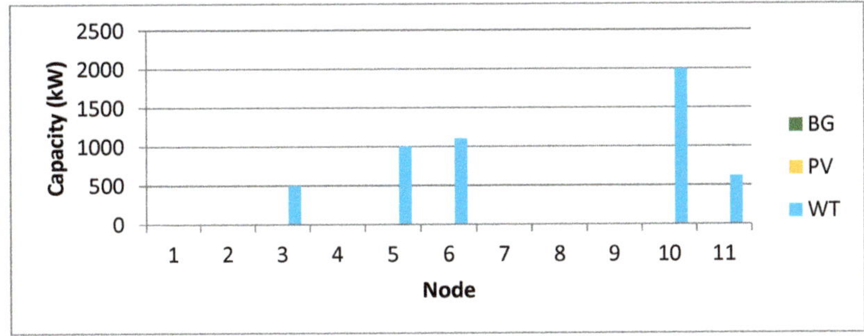

Figure A7. Capacity structure for 30% share of energy from RES in total demand with sizing and allocation of RES and Energy Curtailment (EC).

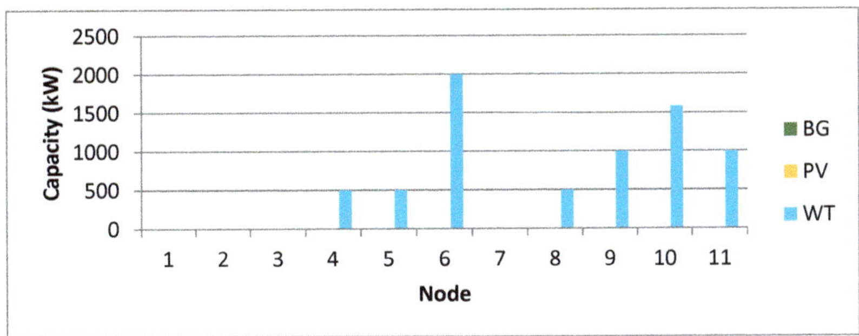

Figure A8. Capacity structure for 40% share of energy from RES in total demand with sizing and allocation of RES and EC.

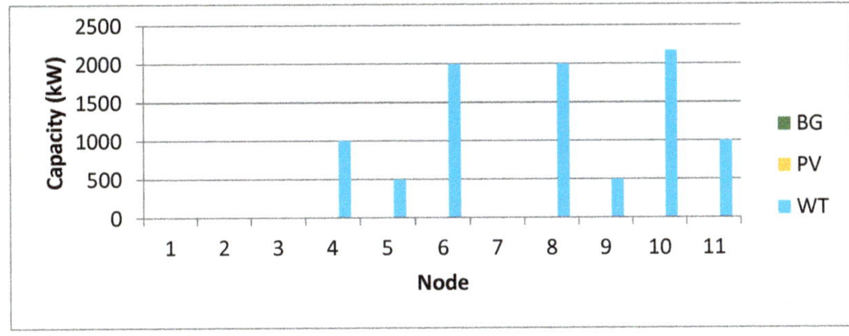

Figure A9. Capacity structure for 50% share of energy from RES in total demand with sizing and allocation of RES and EC.

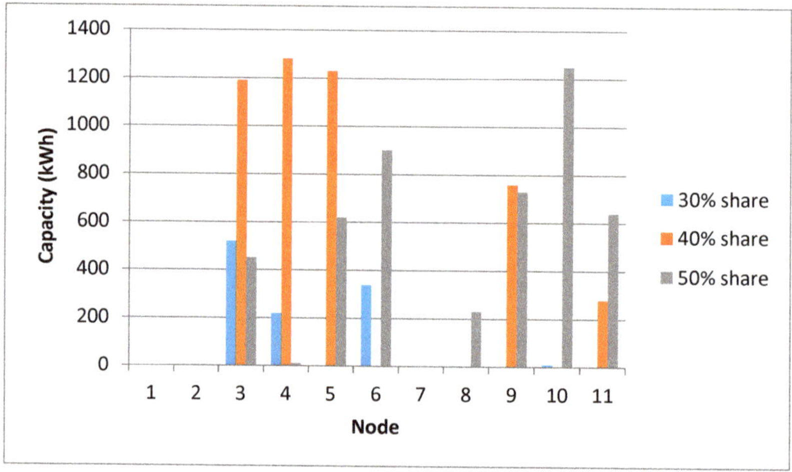

Figure A10. Capacity structure for energy storages in the context of share of energy from RES in total demand.

References

1. European Network of Transmission System Operators for Electricity. *Scenario Outlook & Adequacy Forecast*; ENTSOE: Brussels, Belgium, 2015.
2. Gozel, T.; Hocaoglu, M.H.; Eminoglu, U.; Balikci, A. Optiml placement and sizing of distributed generation on radial feeder with different static load models. In Proceedings of the 2005 International Conference on Future Power Systems, Amsterdam, The Netherland, 18 November 2005; pp. 2–6.
3. Hung, D.Q.; Mithulananthan, N.; Lee, K.Y. Optimal placement of dispatchable and nondispatchable renewable DG units in distribution networks for minimizing energy loss. *Int. J. Electr. Power Energy Syst.* **2014**, *55*, 179–186. [CrossRef]
4. Hung, D.Q.; Mithulananthan, N. Multiple Distributed Generator Placement in Primary Distribution Networks for Loss Reduction. *IEEE Trans. Ind. Electron.* **2013**, *60*, 1700–1708. [CrossRef]
5. Patel, C.A.; Mistry, K.; Roy, R. Loss allocation in radial distribution system with multiple DG placement using TLBO. In Proceedings of the 2015 IEEE International Conference on Electrical, Computer and Communication Technologies (ICECCT), Coimbatore, India, 5–7 March 2015; pp. 1–5.
6. Prakash, C.L.D.B. Multiple DG Placements in Distribution System for Power Loss Reduction Using PSO Algorithm. *Procedia Technol.* **2016**, *25*, 785–792. [CrossRef]
7. Prenc, R.; Škrlec, D.; Komen, V. Distributed generation allocation based on average daily load and power production curves. *Int. J. Electr. Power Energy Syst.* **2013**, *53*, 612–622. [CrossRef]
8. Abu-Mouti, F.S.; El-Hawary, M.E. Optimal Distributed Generation Allocation and Sizing in Distribution Systems via Artificial Bee Colony Algorithm. *IEEE Trans. Power Deliv.* **2011**, *26*, 2090–2101. [CrossRef]
9. Mohandas, N.; Balamurugan, R.; Lakshminarasimman, L. Optimal location and sizing of real power DG units to improve the voltage stability in the distribution system using ABC algorithm united with chaos. *Int. J. Electr. Power Energy Syst.* **2015**, *66*, 41–52. [CrossRef]
10. Babu, P.V.; Dahiya, R. Direct search approach for multiple distributed generator allocation in radial distribution systems. In Proceedings of the 2014 6th IEEE Power India International Conference (PIICON), Delhi, India, 5–7 December 2014; pp. 1–6.
11. Alinejad-Beromi, Y.; Sedighizadeh, M.; Bayat, M.R.; Khodayar, M.E. Using genetic alghoritm for distributed generation allocation to reduce losses and improve voltage profile. In Proceedings of the 2007 42nd International Universities Power Engineering Conference, Brighton, UK, 4–6 September 2007; pp. 954–959.
12. Gopiya Naik, S.N.; Khatod, D.K.; Sharma, M.P. Analytical approach for optimal siting and sizing of distributed generation in radial distribution networks. *IET Gener. Transm. Distrib.* **2015**, *9*, 209–220. [CrossRef]

13. Kumar Injeti, S.; Prema Kumar, N. A novel approach to identify optimal access point and capacity of multiple DGs in a small, medium and large scale radial distribution systems. *Int. J. Electr. Power Energy Syst.* **2013**, *45*, 142–151. [CrossRef]

14. Sultana, S.; Roy, P.K. Multi-objective quasi-oppositional teaching learning based optimization for optimal location of distributed generator in radial distribution systems. *Int. J. Electr. Power Energy Syst.* **2014**, *63*, 534–545. [CrossRef]

15. Bahram, P.; Peyman, K.; Gharehpetian, G.B.; Abedi, M. Optimal allocation and sizing of DG units considering voltage stability, losses and load variations. *Int. J. Electr. Power Energy Syst.* **2016**, *79*, 42–52.

16. Ganguly, S.; Dipanjan, S. Distributed generation allocation with on-load tap changer on radial distribution networks using adaptive genetic algorithm. *Appl. Soft Comput.* **2017**, *59*, 45–67. [CrossRef]

17. Atwa, Y.M.; El-Saadany, E.F. Probabilistic approach for optimal allocation of wind-based distributed generation in distribution systems. *IET Renew. Power Gener.* **2011**, *5*, 79–88. [CrossRef]

18. Bhullar, S.; Ghost, S. Optimal Integration of Multi Distributed Generation Sources in Radial Distribution Networks Using a Hybrid Algorithm. *Energies* **2018**, *11*, 628. [CrossRef]

19. Duong, M.Q.; Pham, T.D.; Nguyen, T.T.; Doan, A.T.; Tran, H.V. Determination of Optimal Location and Sizing of Solar Photovoltaic Distribution Generation Units in Radial Distribution Systems. *Energies* **2019**, *12*, 174. [CrossRef]

20. El-Saadany, E.F.; Abdelsalam, A.A. Probabilistic approach for optimal planning of distributed generators with controlling harmonic distortions. *IET Gener. Transm. Distrib.* **2013**, *7*, 1105–1115.

21. Singh, A.K.; Parida, S.K. Novel sensitivity factors for DG placement based on loss reduction and voltage improvement. *Int. J. Electr. Power Energy Syst.* **2016**, *74*, 453–456. [CrossRef]

22. Gautam, D.; Mithulananthan, N. Optimal DG placement in deregulated electricity market. *Electr. Power Syst. Res.* **2007**, *77*, 1627–1636. [CrossRef]

23. Hadian, A.; Haghifam, M.-R.; Zohrevand, J.; Akhavan-Rezai, E. Probabilistic approach for renewable dg placement in distribution systems with uncertain and time varying loads. In Proceedings of the 2009 IEEE Power & Energy Society General Meeting, Calgary, AB, Canada, 26–30 July 2009; pp. 1–8.

24. Mobin, E.; Mostafa, S.; Masoud, E. Multi-objective optimal reconfiguration and DG (Distributed Generation) power allocation in distribution networks using Big Bang-Big Crunch algorithm considering load uncertainty. *Energy* **2016**, *103*, 86–99.

25. Gil Mena, A.J.; Martín García, J.A. An efficient approach for the siting and sizing problem of distributed generation. *Int. J. Electr. Power Energy Syst.* **2015**, *69*, 167–172. [CrossRef]

26. Doagou-Mojarrad, H.; Gharehpetian, G.B.; Rastegar, H.; Olamaei, J. Optimal placement and sizing of DG (distributed generation) units in distribution networks by novel hybrid evolutionary algorithm. *Energy* **2013**, *54*, 129–138. [CrossRef]

27. Haldar, V.; Chakraborty, N. Reliability based multi-objective DG and capacitor allocation in radial distribution system. In Proceedings of the 2016 IEEE Uttar Pradesh Section International Conference on Electrical, Computer and Electronics Engineering (UPCON), Varanasi, India, 9–11 December 2016; pp. 450–455.

28. Shaaban, M.F.; El-Saadany, E.F. Optimal allocation of renewable DG for reliability improvement and losses reduction. In Proceedings of the 2012 IEEE Power and Energy Society General Meeting, San Diego, CA, USA, 22–26 July 2012; pp. 1–8.

29. Zhu, D.; Broadwater, R.P.; Tam, K.; Seguin, R.; Asgeirsson, H. Impact of DG Placement on Reliability and Efficiency With Time-Varying Loads. *IEEE Trans. Power Syst.* **2006**, *21*, 419–427. [CrossRef]

30. Kim, S.-Y.; Kim, W.-W.; Kim, J.-O. Determining the Optimal Capacity of Renewable Distributed Generation Using Restoration Methods. *IEEE Trans. Power Syst* **2014**, *29*, 2001–2013. [CrossRef]

31. Shaaban, M.F.; Atwa, Y.M.; El-Saadany, E.F. DG allocation for benefit maximization in distribution networks. *IEEE Trans. Power Syst.* **2013**, *28*, 639–649. [CrossRef]

32. Viral, R.; Khatod, D.K. An analytical approach for sizing and siting of DGs in balanced radial distribution networks for loss minimization. *Int. J. Electr. Power Energy Syst.* **2015**, *67*, 191–201. [CrossRef]

33. Mokhlis, H.; Jasmon, G.B.; Aman, M.M.; Bakar, A.H.A. Optimum tie switches allocation and DG placement based on maximisation of system loadability using discrete artificial bee colony algorithm, IET Gener. Transm. Distrib. **2016**, *10*, 2277–2284.

34. Kim, I. Optimal distributed generation allocation for reactive power control, IET Gener. Transm. Distrib. **2017**, *11*, 1549–1556. [CrossRef]

35. Grisales-Noreña, L.F.; Montoya, D.G.; Ramos-Paja, C.A. *Optimal Sizing and Location of Distributed Generators Based on PBIL and PSO Techniques*. *Energies* **2018**, *11*, 1018.
36. Kumar, S.; Mandal, D.P.K.K.; Chakraborty, N. Optimal allocation of multiple DG units in radial distribution system using modified differential evolution technique. In Proceedings of the 2014 International Conference on Control, Instrumentation, Energy and Communication (CIEC), Calcuta, India, 31 January–2 February 2014; pp. 379–383.
37. Gomez-Gonzalez, M.; Hernandez, J.C.; Vera, D.; Jurado, F. Optimal sizing and power schedule in PV household-prosumers for improving PV self-consumption and providing frequency containment reserve. *Energy* **2019**, *191*, 116554. [CrossRef]
38. Babu, P.V.; Singh, S.P. Optimal Placement of DG in Distribution Network for Power Loss Minimization Using NLP & PLS Technique. *Energy Procedia* **2016**, *90*, 441–454.
39. Esmaili, M. Placement of minimum distributed generation units observing power losses and voltage stability with network constraints. *IET Gener. Transm. Distrib.* **2013**, *7*, 813–821. [CrossRef]
40. Bayat, A.; Bagheri, A.; Noroozian, R. Optimal siting and sizing of distributed generation accompanied by reconfiguration of distribution networks for maximum loss reduction by using a new UVDA-based heuristic method. *Int. J. Electr. Power Energy Syst.* **2016**, *77*, 360–371. [CrossRef]
41. Mehdi, R. Simultaneous placement of DG and capacitor in distribution network. *Electr. Power Syst. Res.* **2016**, *131*, 1–10.
42. Rosseti, G.J.S.; de Oliveira, E.J.; de Oliveira, L.W.; Silva, I.C.; Peres, W. Optimal allocation of distributed generation with reconfiguration in electric distribution systems. *Electr. Power Syst. Res.* **2013**, *103*, 178–183. [CrossRef]
43. ti objective placement and sizing of multiple DGs and shunt capacitor banks simultaneously considering load uncertainty via MOPSO approach. *Int. J. Electr. Power Energy Syst.* **2015**, *67*, 336–349. [CrossRef]
44. Carpinelli, G.; Mottola, F. Optimal allocation of dispersed generators, capacitors and distributed energy storage systems in distribution networks. In Proceedings of the 2010 Modern Electric Power Systems, Wroclaw, Poland, 20–22 September 2010; pp. 1–6.
45. Garcia, R.S.; Weisser, D. A wind–diesel system with hydrogen storage: Joint optimisation of design and dispatch. *Renew. Energy* **2006**, *31*, 2296–2320. [CrossRef]
46. Kahrobaee, S.; Asgarpoor, S.; Qiao, W. Optimum sizing of distributed generation and storage capacity in smart households. *IEEE Trans. Smart Grid* **2013**, *4*, 1791–1801. [CrossRef]
47. Delille, G.; Malarange, G.; Gaudin, C. Analysis of the options to reduce the integration costs of renewable generation in the distribution networks. Part 2: A step towards advanced connection studies taking into account the alternatives to grid reinforcement. In Proceedings of the CIRED 22nd International Conference on Electricity Distribution, Stockholm, Sweden, 10–13 June 2013.
48. Pagnetti, A.; Fournel, J.; Santander, C.; Minaud, A. A comparison of different curtailment strategies for distributed generation. In Proceedings of the CIRED 23rd International Conference on Electricity Distribution, Lyon, France, 15–18 June 2018.
49. Tonkoski, R.; Lopes, L.A.C.; El-Fouly, T.H.M. Coordinated Active Power Curtailment of Grid Connected PV Inverters for Overvoltage Prevention. *IEEE Trans. Sust. Energy* **2011**, *2*, 139–147. [CrossRef]
50. Sun, W.; Harrison, G.P. Influence of generator curtailment priority on network hosting capacity. In Proceedings of the CIRED 22nd International Conference on Electricity Distribution, Stockholm, Sweden, 10–13 June 2013.
51. Impact Assesment-Ocena Skutków Regulacji do Projektu Ustawy o Odnawialnych Źródłach Energii. 12 November 2013. Available online: http://www.toe.pl/pl/wybrane-dokumenty/rok-2013?download=72:ocena-skutkow-regulacji (accessed on 12 November 2013).

 *energies*

Article

Maximization of Distribution Network Hosting Capacity through Optimal Grid Reconfiguration and Distributed Generation Capacity Allocation/Control

Rade Čađenović and Damir Jakus *

Department of Power Engineering, University of Split-FESB, 21000 Split, Croatia; Rade.Cadjenovic.00@fesb.hr
* Correspondence: damir.jakus@fesb.hr; Tel.: +385-091-430-5807

Received: 17 August 2020; Accepted: 10 October 2020; Published: 13 October 2020

Abstract: High penetration of small-scale distributed energy sources into the distribution network increase negative impacts related to power quality causing adverse conditions. This paper presents a mathematical model that maximizes distribution network hosting capacity through optimal distributed generation capacity allocation and control and grid reconfiguration. In addition to this, the model includes on-load tap changer control for stabilization of grid voltage conditions primarily in grid operating conditions related to voltage rise problems, which can limit grid hosting capacity. Moreover, the objective function allows the possibility of energy transfer between distribution and transmission grids. The proposed model considers alternative grid connection points for distributed generation and determines optimal connection points as well as install capacity while considering network operating limits. The model is cast as a multiperiod second-order cone linear program and involves aspects of active power management. The model is tested on a modified IEEE 33 bus test network.

Keywords: mathematical programming; small scale distributed generation; distribution networks; active power management

---

1. Introduction

This paper presents a mathematical model for the maximization of small-scale distributed generation (DG) into existing distribution networks (DNs). In multiple cases, the power generated by DG can result in significant operation condition changes of DN causing expensive capital investments in power system equipment and grid. To avoid such scenarios in a manner to facilitate DG penetration into the grid the model maximizes the integration of multiple DGs into the existing DN by using available active power management tools.

The DG connection into DN has become a vital option considering fighting climate changes, a lack of energy worldwide, and power production diversification. Investments in small-scale DG should increase in systems that allow its penetration. The use of traditional methods and tools for DN planning can significantly decrease DN hosting capacity and jeopardize power quality constraints. Hence, high penetration of DG into the DN has to be carefully planned and implemented by using technical and economic planning strategies.

Determination of maximal possible DG penetration into the distribution grids refers to the so-called maximum hosting capacity problem, which can be solved by using optimization tools and different methods. The maximum hosting capacity and DG allocation can be used for solving various problems that are recognized in the distribution system including power loss and energy loss minimization, hosting capacity maximization, voltage profile improvements, reliability enhancement, etc. Authors in [1–4] investigate the possibility of optimal allocation and sizing of DGs to reduce power losses. Photovoltaic (PV) allocation considering optimal grid reconfiguration for power loss reduction is also presented in [5]. In [6], the benefit of DGs allocation is assessed to evaluate the cost of energy losses in DN by using the variable cost of energy. Authors in [7] define the DG allocation optimization problem while considering a multiobjective approach based on cost and benefits related to energy loss costs, system reliability, and cost of purchased energy from transmission grids. Papers with a focus on voltage stability enhancement and loss minimization by taking into account different types of DGs are presented in [8] that are based on the modal analysis and continuation power flow in [9]. In [10] authors presented a practical methodology to determine the dynamic hosting capacity for voltage variations due to power injections and harmonics introduced by distributed energy resources. The authors analyze the change of the levels of harmonic distortions and voltage profile change over a specific time period and its effect on network hosting capacity. Authors in [11] introduced an approach to determine suboptimal energy storage allocation based on voltage profile improvement, which prevents power quality deterioration and allows higher DGs penetration.

Many authors are solving maximum hosting capacity by putting DGs size and allocation as a primary objective function and using the DN flexibility for improving its hosting capacity. In [12] the second-order cone programming method is used for the maximization of PV capacity. In the method that uses genetic algorithms (GAs) optimal hosting capacity is maximized by introducing optimal network reconfiguration. Variation of the DG output is not considered in this method [13]. The heuristic approach is presented in [14] to optimize reconfiguration along with the size and DGs location. A mathematical method based on second-order cone programming is presented in [15]. In this paper, the authors considered topological reconfiguration as the only possible DN flexibility and compare the obtained results with previously fixed topology. The mixed-integer linear programming has been used in [16] for DN reconfiguration with possible connectors for DGs in all DN points. Authors in [17] propose a method for optimal placement DGs by using the loss sensitivity factor method. A multiobjective evolutionary algorithm is proposed by authors in [18] for the sizing and determination of the good locations for DGs by minimizing different functions such as the cost of energy losses, cost of service interruptions, the cost of network upgrading, and the cost of energy purchased. In [19] authors propose a multiperiod AC optimal power flow based technique for maximization of grid hosting capacity, which considers flexibilities such as voltage control and energy curtailment. In [20] the authors introduce streamlined capacity analysis, which provides a fast technique to calculate an impact from small-scale photovoltaics. Static and dynamic reconfiguration is applied in [21] for a multiperiod optimal power flow approach, which is used to assess DGs hosting capacity increment. Authors in [22] propose a maximum hosting capacity evaluation method while considering robust optimal operation and control of on-load tap changer transformers (OLTC) and static VAR compensators.

According to the previous insight in the existing literature, optimization methods for optimal allocation and sizing of DGs are various and include analytical methods, numerical methods, and heuristic methods [23]. Characteristics of analytic methods are their easy implementation and execution, but their results are often only indicative. On the other side numerical methods like nonlinear programming and linear programming usually have better convergence characteristics and can guarantee the determination of global optimum in the case of convex optimization problems. However mathematical programming methods are sometimes not suitable for complex DN and multiperiod programming due to a large number of binary variables in certain approaches. Third, heuristic methods are robust and can provide near optimum results for complex optimization problems [23].

Most of the approaches proposed so far, that address the grid hosting capacity problem, usually consider only a partial set of possible control options. Additionally, these approaches are usually defined as mixed-integer nonlinear optimization problems or they use metaheuristics methods to solve the optimization problem. Given the nature of the optimization problem formulation or the method used for the solution of the problem, such approaches cannot guarantee the detection of a global solution. For example, the approach proposed in [21] cast the grid hosting capacity problem as a mixed-integer nonlinear optimization problem while not considering OLTC or optimal DG allocation in the formulation. Their approach regarding radiality constraints may be insufficient to ensure radiality in grids where there are some zero-injection nodes. Similar goes for the approach used in [16] with the difference that the model is cast as MILP. In addition to that previously mentioned, both of the approaches do not consider multiple operating scenarios derived from real measured data but rather use only the single worst-case scenario for the assessment. Other approaches that were mentioned in the literature overview in addition to the previous do not consider network reconfiguration.

Other approaches that use metaheuristic methods for example [3,4,6], etc. are not suitable to effectively handle the complexity of the model when all control/simulations options (multiscenario, optimal DG allocation and install capacity determination, topology optimization, DG power factor control, and OLTC operation) are taken into account.

From the literature review, it is obvious that there are many methods used for solving the DN hosting capacity problem. When using an approach based on mathematical programming it is very important, like in every other optimization method, to achieve a good convergence rate with low computational requirements. That is why the most mathematical programming models, which are used cannot be implemented in a complex DN and for multiperiod programming. The mathematical programming method based on the SOCP approximations proposed in this paper help to solve some of these issues. Hence, this paper involves effective discretization of time series data (consumption and DG production) by introducing multiperiod fragments thus giving insight into possible operating demand/production scenarios. Significant improvements, in terms of multiperiod optimization and network flexibilities, are achieved by time discretization and by including different network flexibilities (power factor control, OLTC control, and topology reconfiguration) within the hosting capacity optimization problem.

This paper has two significant contributions. First, it gives a possibility for taking into account multiple operating scenarios. This mathematical method is generalized and can be applied for multiperiod optimization (hourly and daily), but this would lead to inconvenient computational time, which is surpassed with scenarios discretization. The other contribution is related to the integration of proposed flexibilities, which allow increment in hosting capacity in existing DN by implementing flexibilities one at a time and combining them into one model. At the end of the paper, mathematical programming models are applied to the modified IEEE 33 bus test case to detect optimal network topology and DG capacity allocation by controlling the DG power factor, OLTC transformer operation, and grid topology configuration. The results are compared with the base model, which does not include any of the DN flexibilities.

2. Mathematical Formulation

The objective function is presented with the following formulation:

$$\text{Minimize} \sum_{i \in B^F,\, s \in S} P_{Ii,s} - \sum_{i \in DI,\, j \in DG_i^{loc},\, s \in S} P_{ij,s}^{DG_{inst}} + \sum_{(ij) \in W,\, s \in S} r_{ij} \frac{p_{ij,s}^2 + q_{ij,s}^2}{v_{i,s}^2} \tag{1}$$

The first term appearing in the formulation of the objective function represents the possibility of the export of the energy between the interconnected grids. The second term represents the total install capacity of distributed generation and the third is related to active power losses. The objective function is followed by the set of constraints defined as:

1.   Bus active and reactive power balance:

$$\sum_{i:(j,i)\in W} p_{ij,s} = \sum_{(ij)\in W,\ s\in S} r_{ij}\frac{p_{ij,s}^2+q_{ij,s}^2}{v_{i,s}^2} + P_{j,s}^L - P_{j,s}^{gen} - P_{j,s}^{DG_{inst}} + \sum_{k:(jk)\in W,\ s\in S} p_{jk,s},\quad j\in B\backslash B^F$$

$$P_{j,s}^{gen} = \sum_{k:(jk)\in W} p_{jk,s} + P_{j,s}^L - P_{j,s}^{DG_{inst}},\quad j\in B^F$$

$$\sum_{i:(j,i)\in W} q_{ij,s} = \sum_{(ij)\in W,\ s\in S} x_{ij}\frac{p_{ij,s}^2+q_{ij,s}^2}{v_{i,s}^2} + Q_{j,s}^L - Q_{j,s}^{gen} - Q_{j,s}^{DI_{inst}} + \sum_{k:(jk)\in W,\ s\in S} q_{jk,s},\quad j\in B\backslash B^F$$

$$Q_{j,s}^{gen} = \sum_{k:(jk)\in W} q_{jk,s} + Q_{j,s}^L - Q_{j,s}^{DI_{inst}},\quad j\in B^F$$

(2)

2. Branch voltage drop (higher-order terms in the expression for branch voltage drop are left out to linearize the expression):

$$v_{j,s}^2 - v_{i,s}^2 \le r_{ij}\left(p_{ji,s}-p_{ij,s}\right) + x_{ij}\left(q_{ji,s}-q_{ij,s}\right) + M\left(1-y_{ij}\right),\quad i,j\in W$$

$$v_{j,s}^2 - v_{i,s}^2 \ge r_{ij}\left(p_{ji,s}-p_{ij,s}\right) + x_{ij}\left(q_{ji,s}-q_{ij,s}\right) - M\left(1-y_{ij}\right),\quad i,j\in W$$

(3)

3. Radial network constraints [24]:

$$z_{ij} \ge 0$$
$$z_{if} = 0,\quad f\in B^F$$
$$z_{ij} + z_{ji} = 1,\quad (i,j)\in W\backslash W^S$$
$$z_{ij} + z_{ji} = y_{ij},\quad (i,j)\in W^S$$
$$\sum_{j:(i,j)\in W} z_{ji} = 1,\quad i\in B\backslash B^F$$
$$y_{ij} \in \{0,1\},\quad (i,j)\in W^S$$

(4)

In the distribution networks, a large number of switching operations are even today done manually, given that most of the distribution network switchgear is not fully automatized and centrally controlled due to a lack of underlying communication infrastructure. In these networks, dynamic network reconfiguration is not an option given that every topology change would require a significant amount of time leaving certain consumers out of the operation. Given this, in the proposed model we did not consider dynamic network reconfiguration but a rather static reconfiguration. This means that we determined optimal network topology, which maximizes network hosting capacity, and this topology was kept constant in all considered operating scenarios. Optimal network topology was determined while also optimizing DG connection point and capacity. Modification of the proposed model to include dynamic network reconfiguration is a rather straightforward process that would additionally increase grid hosting capacity but probably not in the amount to apply such a complex operating procedure.

Given this, the proposed approach does not result in frequent network switchgear operation. Other control parameters considered in this paper, OLTC ratio and DG power factor control vary with load and DG production variations.

4. DG connection and capacity constraints:

$$\sum_{j\in DI_i^{lok}} y_{i,j}^{DG} \le 1,\quad \forall i\in DG$$

(5)

$$\sum_{j \in DG_i^{loc}} P_{i,j}^{DG_{inst}} \le P_i^{DG_{max}}, \quad \forall i \in DG$$

$$0 \le P_{i,j}^{DG_{inst}} \le y_{i,j}^{DG} \cdot P_i^{DG_{max}}, \quad \forall \left(i \in DG, \ j \in DG_i^{loc} \right) \tag{6}$$

In real distribution networks, DSOs usually cannot force grid connection location upon *DG* investors to improve voltage stability if the *DG* unit is not the reason for the voltage stability problem. The *DG* investor, in the grid connection study, usually considers a couple of grid connection options and usually chooses the cheapest solution to maximize the profit. In the proposed mathematical model we define upfront the possible grid connection points for each *DG* unit separately. The number of possible grid connection points is arbitrary and the model allows one to consider each system bus as a potential connection point. Given that we simultaneously tried to maximize *DG* penetration, the model determines the optimal grid connection point and installed capacity. The model also reconfigures the network topology. Using these approaches, weak (critical) buses are correctly addressed with the model automatically because the algorithm will automatically "reshape" the network to tackle low voltage problems in parts of the grid due to high load or to tackle high voltage problems due to *DG* production. This is done through network reconfiguration (in the network segments, which are meshed) and optimal *DG* allocation in combination with OLTC control and *DG* power factor control.

5. *DG* power factor constraints:

$$Q_{i,j,s}^{DG_{inst}} = tg(\varphi)_{i,j,s}^{DG_{inst}} \cdot P_{i,j,s}^{DG_{inst}}, \quad \forall \left(i \in DG, \ j \in DG_i^{loc} \right)$$

$$\left| tg(arccos(\varphi))_{i,j,s}^{DG_{inst}} \right| \le 0.326 \tag{7}$$

6. On-load tap changer transformer constraints located in the interconnection substation [25]:

$$tr_s = tap_{min} + n_s \cdot \Delta tap, \quad 0 \le n_s \le n_{s,max}, \quad n_s \in integer$$

$$\Delta tap = (tap_{max} - tap_{min}) / tap_{max} \tag{8}$$

$$v_{1,s}^2 = tr_s^2 \cdot v_{0,s}^2, \qquad v_{0,s}^2 = const.$$

The OLTC is an important and very expensive part of a power transformer and the main cause of power transformer failures. The OLTC probability failure can be directly linked to the number of switching operations so the frequent tap change operation significantly reduces the component lifetime. Given that the proposed model approximates real time-series data with a set of representative operating scenarios that are not inter temporally linked, it is not possible to restrict or assess the frequency of tap changer operation and its effect on component lifetime. Once the proposed model determines optimal DG connection points and capacity and optimal network topology, an additional optimization model that optimizes network operation (losses reduction and voltage profile improvement) could be used to assess and limit the frequency of the tap changer operation. In such a model the limitations regarding the frequency of tap changers could be explicitly stated. A previous study regarding this matter indicates that the slow voltage changes could be handled without the too frequent operation of OLTC given that most *DG* units are also capable of providing voltage support usually in the power factor range 0.95 leading/lagging. Additionally, voltage support requirement is initially reduced by network topology reconfiguration, which will reorganize load and production across feeders to improve element loading and bus voltages while considering the defined operational scenario set.

where:

- *B*—set of network buses,
- B^F—set of supply network buses,
- *W*—set of network lines,

- W^S—set of network lines available for reconfiguration,
- s—set of available scenarios,
- DG—set of distributed generators,
- DG_i^{loc}—set of potential connection buses of distributed generator i,
- $P_{i,s}^L / Q_{i,s}^L$—active/reactive load within scenario s at bus i,
- $P_{i,j,s}^{DI_{inst}}$—active power generation of distributed generator i at bus j within scenario s,
- $Q_{i,j,s}^{DI_{inst}}$—reactive power generation of distributed generator i at bus j within scenario s,
- $P_i^{DI_{maks}}$—maximum capacity of i-th distributed generator,
- $P_{i,s}^{gen} / Q_{i,s}^{gen}$—active/reactive power of supply point i within scenario s,
- $y_{i,j}^{DG}$—discrete variable indicating the connection status of the i-th distributed generator to bus j,
- $p_{ij,s}$—active power flow for line $i-j$ within scenario s,
- $v_{0,s}$—supply point voltage of substation i within scenario s,
- $v_{i,s}$—the voltage at bus i within scenario s,
- r_{ij}—the resistance of line $i-j$,
- x_{ij}—the reactance of line $i-j$,
- $q_{ij,s}$—reactive power flow for line $i-j$ within scenario s,
- M—a large number,
- y_{ij}—discrete switch variable for line $i-j$,
- z_{ij}—continuous orientation variable for line $i-j$,
- $cos(\varphi)_{i,j,s}^{DG_{inst}}$—distributed generator i power factor at bus j within scenario s,
- tap_{min} / tap_{maks}—minimal/maximal transformer ratio,
- Δtap—OLTC transformer tap change,
- n_s—tap change position of OLTC transformer,
- $n_{s,max}$—maximal position of tap change,
- tr_s—transformer ratio within scenario s,
- s_{ij}^{max}—power capacity of the line $i-j$.

2.1. Second-Order Cone Programming (SOCP) Approximations

The SOCP approximations aim to transform the nonlinear expressions, which are present in Equations (1) and (2) by the following substitutions:

$$L_{ij,s} = \frac{p_{ij,s}^2 + q_{ij,s}^2}{v_{i,s}^2}, \quad (i,\ j) \in W \backslash W^S$$

$$L_{ij,s} = \frac{p_{ji,s}^2 + q_{ji,s}^2}{v_{j,s}^2}, \quad (i,\ j) \in W \backslash W^S \tag{9}$$

$$u_{j,s} = v_{i,s}^2, \quad i \in B$$

Variables defined within equation set (9) are replaced into equation sets (2) and (3):

$$\sum_{i:(ji)\in W} p_{ij,s} = \sum_{(ij)\in W,\ s\in S} r_{ij} \cdot L_{ij,s} + P_{j,s}^L - P_{j,s}^{gen} - P_{j,s}^{DG_{inst}} + \sum_{k:(jk)\in W,\ s\in S} p_{jk}, \quad j \in B \backslash B^F$$

$$\sum_{i:(ji)\in W} q_{ij,s} = \sum_{(ij)\in W,\ s\in S} x_{ij} \cdot L_{ij,s} + Q_{j,s}^L - Q_{j,s}^{gen} - Q_{j,s}^{DG_{inst}} + \sum_{k:(jk)\in W,\ s\in S} q_{jk}, \quad j \in B \backslash B^F \tag{10}$$

$$u_{j,s} - u_{i,s} \leq r_{ij}\left(p_{ji,s} - p_{ij,s}\right) + x_{ij}\left(q_{ji,s} - q_{ij,s}\right) + M\left(1 - y_{ij}\right), \quad i,j \in W$$

$$u_{j,s} - u_{i,s} \geq r_{ij}\left(p_{ji,s} - p_{ij,s}\right) + x_{ij}\left(q_{ji,s} - q_{ij,s}\right) - M\left(1 - y_{ij}\right), \quad i,j \in W \tag{11}$$

Furthermore, SOCP constraints are given with a set of formulation below:

$$L_{ij,s} \geq \frac{p_{ij,s}^2 + q_{ij,s}^2}{v_{i,s}^2 = u_{i,s}} \quad => \quad \left\| \begin{array}{c} 2p_{ij,s} \\ 2q_{ij,s} \\ L_{ij,s} - u_{i,s} \end{array} \right\|_2 \leq L_{ij,s} + u_i \quad => \quad p_{ij,s}^2 + q_{ij,s}^2 \leq L_{ij,s} \cdot u_{i,s}$$

$$L_{ij,s} \geq \frac{p_{ji,s}^2 + q_{ji,s}^2}{v_{j,s}^2 = u_{j,s}} \quad => \quad \left\| \begin{array}{c} 2p_{ji,s} \\ 2q_{ji,s} \\ L_{ij,s} - u_{j,s} \end{array} \right\|_2 \leq L_{ij,s} + u_{j,s} \quad => \quad p_{ji,s}^2 + q_{ji,s}^2 \leq L_{ij,s} \cdot u_{j,s} \tag{12}$$

Additional constraints are introduced to reflect restrictions related to power line capacity and bus voltage limits:

$$L_{ij,s} \leq y_{ij} \cdot s_{ij}^{max}, \quad i,j \in W \tag{13}$$

$$u_i = \left(v_i^{set}\right)^2, \quad i \in B^F$$

$$\left(v_i^{MIN}\right)^2 \leq u_{i,s} \leq \left(v_i^{MAX}\right)^2, \quad i \in B \backslash B^F \tag{14}$$

where:

- s_{ij}^{max} —maximal power capacity of the line $i - j$ in p.u.,
- v_i^{MIN} / v_i^{MAX} —minimal/maximal voltage at bus i.

This paper did not consider voltage improvements as the primary objective function but includes the bus voltage constraints to ensure the required voltage profile and power quality (Equations (11) and (14)) under a different set of operating scenarios. The voltage constraints are modeled as hard constraints that maintain normal voltage conditions for all considered operating scenarios regarding the different network load and DG production levels. The proposed model does not consider power quality issues related to harmonics injected by DG units and power electronics at grid interfaces. This factor can also be a limiting factor especially in weak distribution networks in which THD can be significantly affected by DG connection. To tackle THD and limit the levels at each system bus according to power quality standards, the model should include harmonic power flow calculation, which would complicate the model and significantly prolong the computational time.

2.2. Model Linearization of OLTC Transformer

Nonlinear OLTC transformer constraints formulated in (8) can be linearized by introducing additional binary variables [25]. To linearize the OLTC model, we introduced the binary encoding of a tap position in the following manner:

$$n_s = \sum_{a=0}^{bin} 2^a \cdot \tau_{a,s} \tag{15}$$

where we had a new binary variable $\tau_{a,s}$. Given this, the OLTC transformer model in (8) can be reformulated into Equation (14):

$$tr_s = tap_{min} + \Delta tap \sum_{a=0}^{bin} 2^a \cdot \tau_{a,s}, \quad \sum_{a=0}^{bin} 2^a \tau_{a,s} \leq tap_{max}, \tag{16}$$

$$\tau_{a,s} \in \{0,1\}, \quad \forall\, a \in \{0, \ldots, bin\}$$

Considering the previous formulation and by introducing a new variable $\rho_{a,s} = \tau_{a,s} \cdot tr_{a,s}$, $\forall\, a \in \{0,\dots,bin\}$ and big number S, voltage formulation within OLTC transformer substation is defined as:

$$u_{1,s} = u_{0,s} \cdot tap_{min} \cdot tr_s + u_{0,s} \cdot \Delta tap \sum_{a=0}^{bin} 2^a \cdot \rho_{a,s}, \qquad u_{0,s} = const.$$

$$0 \le tr_s - \rho_{a,s} \le (1 - \tau_{a,s})S \tag{17}$$

$$0 \le \rho_{a,s} \le \tau_{a,s}S, \quad \forall\, a \in \{0,\dots,bin\}, \quad S - big\ number$$

By these means, we obtained a linear mathematical model of the OLTC transformer. Where:

- S—a big number,
- $\tau_{a,s}$—binary variable describing the position of tap change of OLTC transformer,
- bin—length of the binary representation of tap change position of OLTC transformer,
- $\rho_{a,s}$—additional variable introduced to linearize nonlinear expression incurred by multiplying two variables.

2.3. Mixed-Integer Second-Order Cone Programming (MISOCP) Model

After the introduction of SOCP approximations and linearization of the OLTC transformer model, the complete set of equations used in this model is represented as follows:

1. By introducing (9) into (1), the objective function is given as:

$$Minimize \sum_{i \in B^F,\, s \in S} P_{Ii,s} - \sum_{i \in DI, j \in DI_i^{lok},\, s \in S} P_{ij,s}^{DI_{inst}} + \sum_{(ij) \in W, s \in S} r_{ij} \cdot L_{ij,s} \tag{18}$$

   subject to:
2. Node active and reactive power balance constraints given by the equation set (10).
3. Power flow, line current equations given by (12) and (13)
4. Voltage constraints defined by equation sets (11) and (14),
5. Radial constraints defined by equation set (4),
6. DG location constraints, capacity constraints, and power factor constraints defined by equations sets (5), (6), and (7),
7. OLTC transformer model constraints given with equation sets (16) and (17).

3. Case Study

3.1. Multiperiod Simulation and Scenario Selection

In most of the approaches, the maximum grid hosting capacity is determined while analyzing the worst-case scenario or by selecting a few representative scenarios of DG production/load consumption. This approach usually does not cover all possible operation scenarios, nor does it account for the probability or duration of scenarios.

In the proposed approach, the objective is to maintain a normal system operating state for the entire set of operating states considered in the model. Therefore, the input data used model is generalized considering the possibility of obtaining multiple operating conditions based on demand/production variations. Furthermore, this analysis does not take into account dynamic reconfiguration but rather gives a unique static topology solution, which is applied for all considered operation scenario set.

The input data for this case study is the time-series data of network load and DG production, which accounts for the correlation between these relevant parameters. Using this raw data in the optimization problem would introduce a large number of variables, which would result in high computational time. On the other hand, limiting the number of scenarios to only a few extreme cases does not account for the probability of a system operating state. The approach proposed here constructs blocks of scenarios using available time-series data based on a load/DG generation duration curves, which are constructed from available time-series data. Simulation is conducted based on real measured data related to wind power plant production, sun intensity, and network demand.

In the case study, we considered a few optimization models based on the level of the operational flexibilities, which are considered in the models. These submodels that do not consider all previously mentioned flexibilities could be easily derived from the proposed full optimization model by fixing the values of certain variables. Models implemented for comparison are defined as follows:

- Base model—this submodel does not include any of the DN flexibilities stated in the proposed mathematical model. The model uses initial network radial topology, DG units operate with the unit power factor, and OLTC is fixed in a central position. Within this analysis, all the DN system flexibilities are excluded.
- Model (a)—this submodel considers DGs power factor flexibility.
- Model (b)—this submodel combines two flexibilities: DG power factor control and OLTC contribution for the solution of voltage rise/drop problem.
- Model (c)—this case includes all three possible flexibilities: DN reconfiguration, power factor regulation, and OLTC transformer control.

The initial parameters of the simulation are given in Table 1.

Table 1. Simulation parameters.

Power Factor Range	$cos(\varphi)=0.95(inductive/capacitive)$	
Power Line Ratings	Lines 1–17=$10\,MVA$	Lines 18–37=$5\,MVA$
OLTC number of tap changers/tap change Bus voltages range	20/1% 0.9–$1.1\,p.\,u.$	

One of the most used network models for distribution grid analysis and simulation is the IEEE 33 bus test network. The network consists of 37 branches and 32 demand buses with one supply point. It contains 5 elementary loops and is presented in Table 2 [26]. The offline branches indicated in Table 2 are related to the initial network radial topology. In this case study, we assumed that all branches are available for network reconfiguration if they are part of the network elementary cycles.

Table 2. IEEE 33 bus test network.

Test Network	Initial Network Topology	
	Offline Branches	$P_{load\_max},\ Q_{load\_max}$
IEEE 33 bus	33, 34, 35, 36, 37	3715 kW, 2300 kVA

The full meshed modified IEEE 33 bus test network, with all possible DG connection points, is shown in Figure 1. The maximum install capacity of all DG units (wind and solar) is 10 MW and for every DG unit, only one connection point can be realized.

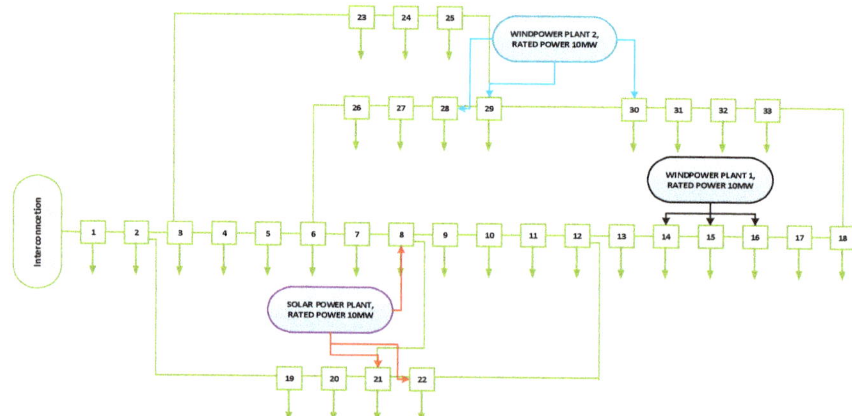

Figure 1. IEEE 33 bus test network with potential distributed generation (DG) connection buses.

Figure 2 shows the wind power plant (WPP) and PV plant relative production for one year with an hourly resolution. This real time-series data together with network consumption data was used to construct a representative scenario set of operating scenarios that were used in the proposed model for maximization of network hosting capacity.

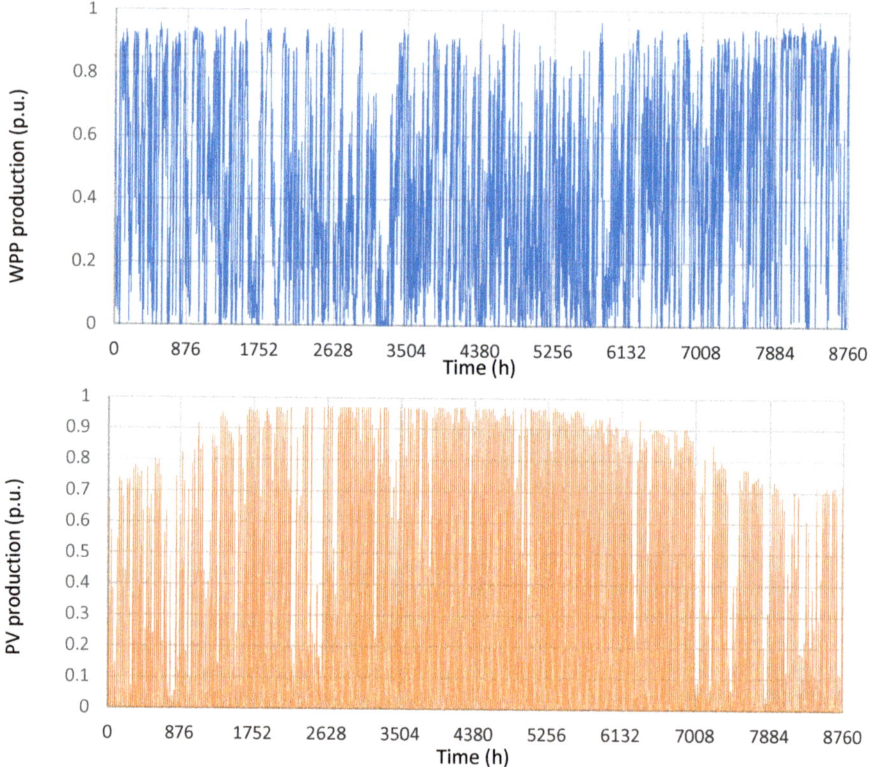

Figure 2. Wind power plant (WPP) and photovoltaic (PV) plant relative production.

The available time-series data related to network consumption and normalized DG production is represented by its load/production duration curves [27]. The load duration curve was used as a key for sorting DG production duration curves. In order to reduce the model computational burden, the load duration curve, solar power curve and wind power curve were divided into 4 demand/production blocks as shown in Figure 3. The demand variations within each demand block were approximated by a set of scenarios, namely, high, average, and low. The DG production duration curve associated with each demand block was also approximated with a set of 3 scenarios, namely, high, average, and low. This way, originally measured time-series data was approximated by jointly considering the demand and DG levels, which in turn results with a representative set of 36 operating scenarios: three demand levels by three DG levels by four demand blocks (Table 3). Using this clustering technique it is possible to maintain a correlation between network consumption and DG production data while significantly reducing the model computational burden. The number of levels, and blocks, can be differently defined to achieve a better approximation of original time-series data at the expense of computational time.

Table 3. Parameters for 36 operating scenarios considered in the analysis.

Scenario	Block	Hour (h)	Load (p.u.)	Wind/Solar Production (p.u.)
1			0.9429	0.938/0.915
2			0.9429	0.5735/0.596
3			0.9429	0.2443/0
4			0.7011	0.938/0.915
5	1	1200	0.7011	0.5735/0.596
6			0.7011	0.2443/0
7			0.52	0.938/0.915
8			0.52	0.5735/0.596
9			0.52	0.2443/0
10			0.52	0.927/0.903
11			0.52	0.532/0.533
12			0.52	0.1834/0
13			0.4628	0.927/0.903
14	2	3600	0.4628	0.532/0.533
15			0.4628	0.1834/0
16			0.41795	0.927/0.903
17			0.41795	0.532/0.533
18			0.41795	0.1834/0
19			0.389	0.885/0.886
20			0.389	0.47/0.488
21			0.389	0.1247/0
22			0.341	0.885/0.886
23	3	2400	0.341	0.47/0.488
24			0.341	0.1247/0
25			0.2718	0.885/0.886
26			0.2718	0.47/0.488
27			0.2718	0.1247/0
28			0.2711	0.884/0.71
29			0.2711	0.43/0.276
30			0.2711	0.117/0
31			0.223	0.9045/0.71
32	4	1560	0.223	0.43/0.276
33			0.223	0.117/0
34			0.19	0.9045/0.71
35			0.19	0.43/0.276
36			0.19	0.117/0

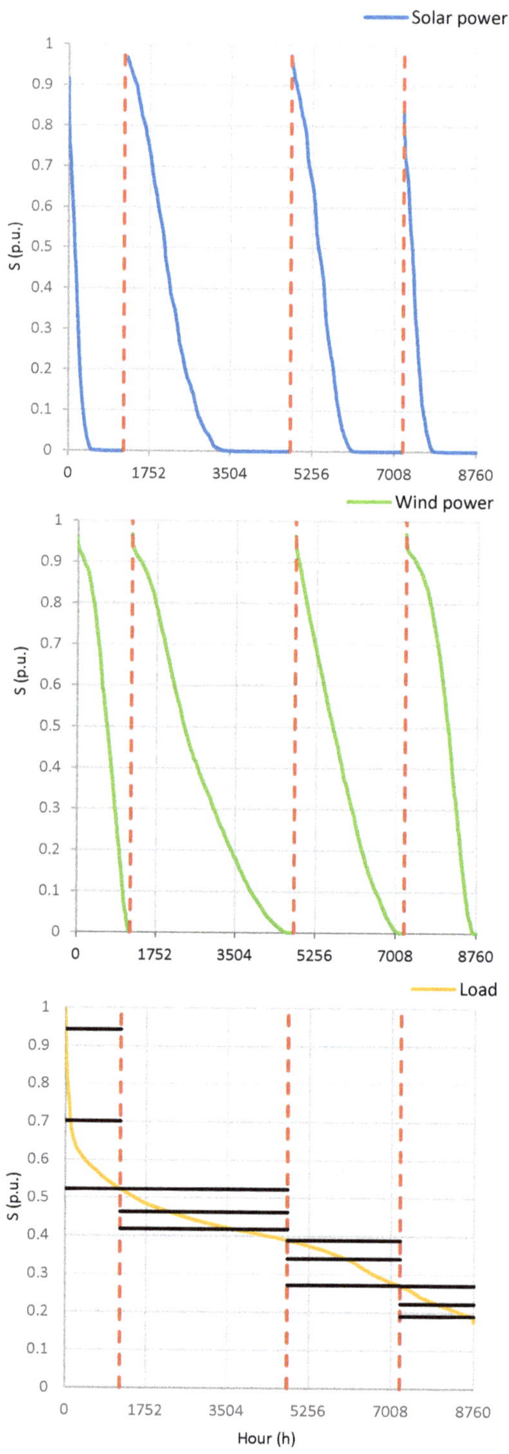

Figure 3. Load consumption—DG production duration curves divided into 4 blocks.

3.2. Implementation of the Mathematical Model and Results Discussion

To assess grid hosting capacity based on upper data settings, and provide adequate numerical results, we implemented a mixed-integer second-order cone programming (MISOCP) mathematical model described in the previous chapter in the general algebraic modeling system (GAMS) [28] and solved underlying problems using CPLEX software (IBM, Armonk, NY, USA) [29]. Tests were performed on a Windows machine equipped with an Intel Core i3 (2.27 GHz) processor and 4 GB of RAM. By implementing a mathematical model we obtained results that are presented in Table 4.

Table 4. Grid hosting capacity and optimal DG allocation and capacity results.

Model	Base Model	Model (a)	Model (b)	Model (c)
Rated Power of WPP1, WPP2, PV (MW)	1.54, 4.019 4.884	2.046, 5.29 35.596	2.384, 5.865 5.501	3.07, 5.885 5.317
Increment (MW)	-	0.524, 1.274 0.712	0.844, 1.846, 0.617	1.53, 1.889 0.433
Connection bus for WPP1, WPP2, PV	15, 28, 21	15, 28, 21	15, 28, 21	15, 29, 21
Offline branches	33, 34, 35, 36, 37	33, 34, 35, 36, 37	33, 34, 35, 36, 37	9, 16, 21, 25, 33

WPP—wind power plant, PVP—photovoltaic power plant.

We can see from the results shown in Table 4 that considered submodels produce different results related to optimal DG allocation and capacity.

The base model, which did not include any of the considered flexibilities and was simulated for 36 different operating scenarios, reached a total hosting level of 10.444 MW. According to the results of the base model, the total hosting capacity was distributed as follows: WPP1 optimal install capacity was 1.54 MW with connection to bus 15, WPP2 optimal install capacity was 4.019 MW with connection to bus 28, and PVP optimal install capacity was 4.884 MW with connection to bus 21. Further increment of DG power, in this case, would lead to violation of DN operational constraints (bus voltage constraints would be violated in certain operating scenarios). This total hosting capacity represents the referent value for comparison with other submodels.

Model "a" considers the possibility of a DG power factor control in the range $\cos\varphi = 0.95$ (leading/lagging), which leads to an increment of network hosting capacity to 12.935 MW with the highest increment of install power for WPP2 (optimal installed power increased by 1.274 MW). Model "b", in addition to DG power factor control, considers the possibility of OLTC voltage control. Based on this submodel the limits for DG penetration were additionally increased to 13.75 MW, with the highest increase of install capacity for WPP2. Furthermore, in the submodel "c" we included the possibility of network reconfiguration in addition to DG power factor and OLTC control. By redistributing power from DGs through optimal power network topology modification while maintaining radial network operation, further increase of network hosting capacity can be achieved. To increase network hosting capacity the model suggests topology modification by switching offline branches 9, 16, 21, 25, and 33. In this case, we reached the maximal level of DG integration equal to 14.272 MW, making this approach the most convenient for the maximization of network hosting capacity. This level of DG penetration represents an increment of 37% compared to the base model.

Total network active power losses for different operating scenarios are shown in Figure 4. The analysis shows an increase in power losses due to DG power penetration. It is interesting to note that network power losses were lower for submodel "c" in comparison with submodels "a" and "b" although total install DG capacity was higher. The reason for this is network topology optimization included in submodel "c", which not only did it increase network hosting capacity but it also reduced network active power losses.

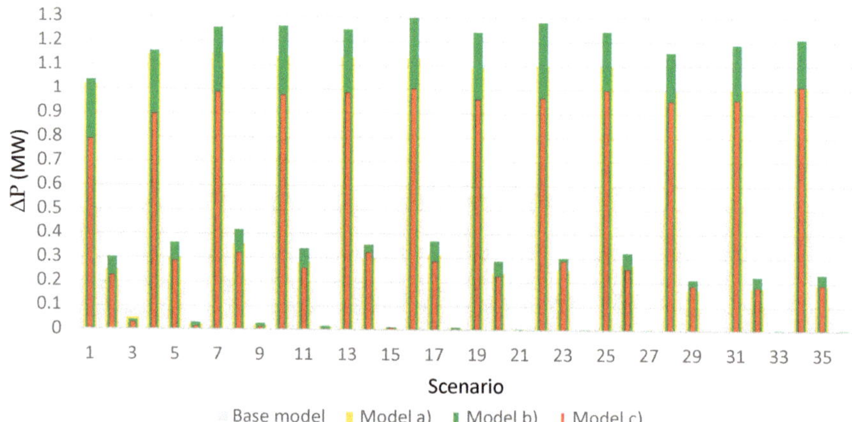

Figure 4. Total power losses for different operating scenarios and submodels.

Figure 5 shows minimum/maximum bus voltage ranges for four different submodels and all considered operating scenarios. It is evident that the voltage level stayed within the limits in all operating scenarios. Moreover, the limiting factor for a further increment of network hosting capacity was visible from the figures and was directly related to the increase of the voltage in DG connection buses and adjacent grid at least in one scenario included in the simulation.

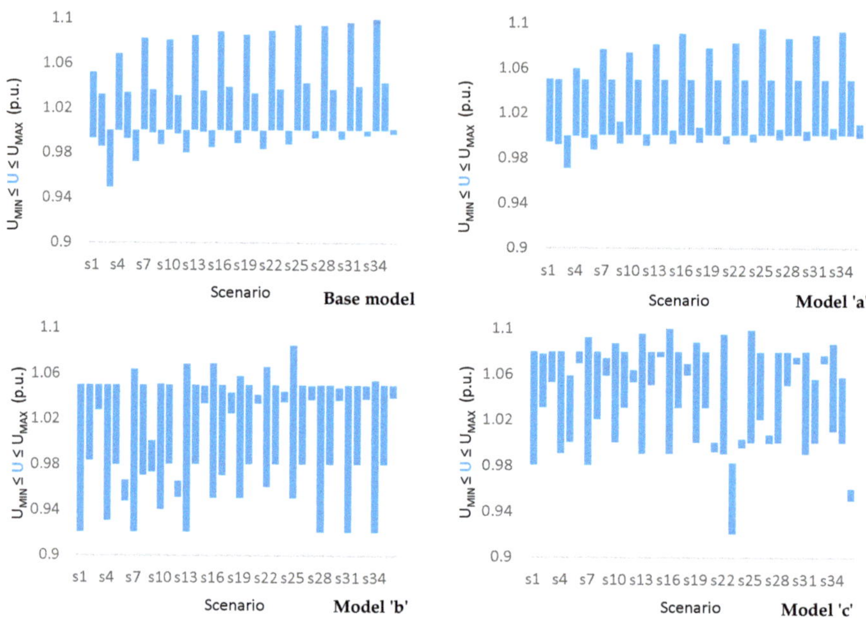

Figure 5. Voltage ranges for different operating scenarios and submodels.

Figure 6 shows a comparison of maximal line loadings for power lines '1–17' (maximal power rating 10 MVA) and in Figure 7 for lines '18–37' (maximal power rating 5 MVA) for all considered operating scenarios and four submodels. It is evident that upper power ratings of the lines were not surpassed but rather within the element power rating limits. Looking at the bus voltage results in parallel with the results of line loading it is obvious that in the base model main limitation factor for DG penetration was not connected with line overloading but rather to the voltage rise problem. Models "a", "b", and "c" on the other hand show limitations related both to voltage rise problems and line loading problems. In these models, upper loading limits of the power lines were reached making additional barriers along with the voltage level for further DG power penetration.

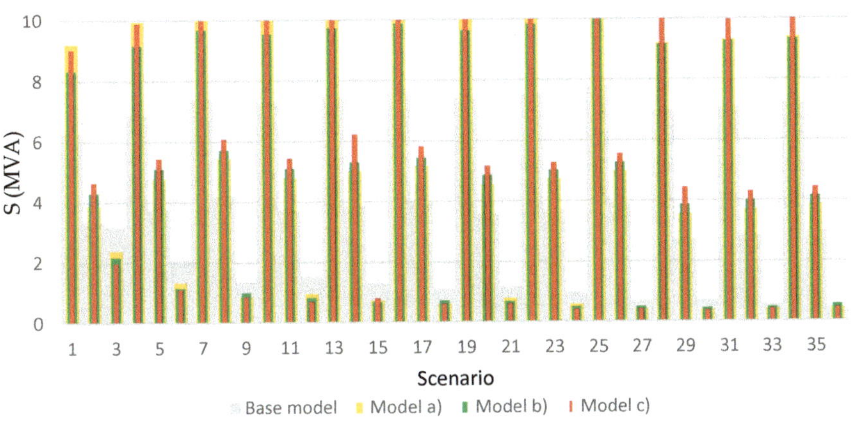

Figure 6. Maximal power line loading (branches '1–17') for different operating scenarios and submodels.

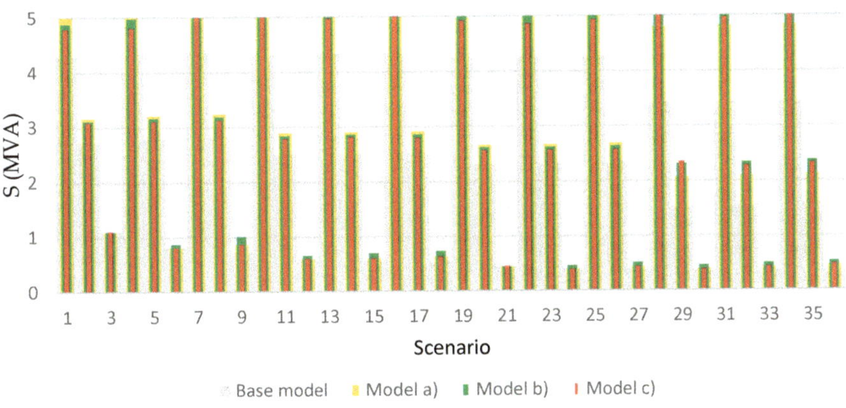

Figure 7. Maximal power line loadings (branches '18–37') for different operating scenarios and submodels.

4. Conclusions

This paper proposed a MISOCP mathematical model for the multiperiod maximization of DGs penetration into the existing DN. The model includes different network flexibilities such as the DG power factor control, OLTC transformer control, and network topology reconfiguration. The proposed approach includes the discretization of demand and DG production duration curves and the construction of a representative operating scenario set. This approach accounts for the correlation between relevant system variables and at the same time considers fewer scenarios thus reducing the level of operating conditions to a level appropriate for mathematical modeling optimization. The proposed method was tested on a modified IEEE 33 bus test network. Modifications refer to three DGs added to the grid with a presumed maximum rated power and defined set of potential grid connection points. Simulation analysis covered four different submodels and compared results related to maximum grid hosting capacity and optimal DG allocation and installed capacity. The model that includes network topological flexibility gave a maximal increment of grid hosting capacity in relation to the base submodel. The other submodels that did not include topological flexibility were affected with line overloading and hence allowed lower grid hosting capacity. Results show a significant increment in network hosting capacity in all three submodels compared to the base model, with the largest increase of grid hosting capacity in a model that considered network topology reconfiguration, DG power factor control, and OLTC voltage control with a 37% increase of grid hosting capacity in relation to the base case model.

Author Contributions: D.J. and R.Č. defined the main concepts and implemented the proposed algorithms. All authors have read and agreed to the published version of the manuscript.

Funding: This research received no external funding.

Conflicts of Interest: The authors declare no conflict of interest.

References

1. Atwa, Y.M.; El-Saadany, E.F.; Salama, M.M.A.; Seethapathy, R. Optimal Renewable Resources Mix for Distribution System Energy Loss Minimization. *IEEE Trans. Power Syst.* **2009**, *25*, 360–370. [CrossRef]
2. Griffin, T.; Tomsovic, K.; Secrest, D.; Law, A. Placement of dispersed generation systems for reduced losses. In Proceedings of the 33rd Annual Hawaii International Conference on System Sciences (HICSS), Maui, HI, USA, 7 January 2000; pp. 1446–1454.
3. Vatani, M.; Gharehpetian, G.B.; Sanjari, M.J.; Alkaran, D.S. Multiple distributed generation units allocation in distribution network for loss reduction based on a combination of analytical and genetic algorithm methods. *IET Gener. Transm. Distrib.* **2016**, *10*, 66–72. [CrossRef]
4. Pal, A.; Chakraborty, A.K.; Bhowmik, A.R.; Bhattacharya, B. Optimal DG allocation for minimizing active power loss with better computational speed and high accuracy. In Proceedings of the 2018 4th International Conference on Recent Advances in Information Technology (RAIT), Dhanbad, India, 21 June 2018; pp. 1–6.
5. Bouhouras, A.S.; Papadopoulos, T.A.; Christoforidis, G.C.; Papagiannis, G.K.; Labridis, D.P. Loss reduction via network reconfigurations in Distribution Networks with Photovoltaic Units Installed. In Proceedings of the 2013 10th International Conference on the European Energy Market (EEM), Stockholm, Sweden, 26 September 2013; pp. 1–8.
6. Shaaban, M.F.; Atwa, Y.M.; El-Saadany, E.F. DG allocation for benefit maximization in distribution networks. *IEEE Trans. Power Syst.* **2012**, *28*, 639–649. [CrossRef]
7. Khalesi, N.; Rezaei, N.; Haghifam, M.-R. DG allocation with application of dynamic programming for loss reduction and reliability improvement. *Int. J. Electr. Power Energy Syst.* **2011**, *33*, 288–295. [CrossRef]
8. Mehta, P.; Bhatt, P.; Pandya, V. Optimal selection of distributed generating units and its placement for voltage stability enhancement and energy loss minimization. *Ain Shams Eng. J.* **2018**, *9*, 187–201. [CrossRef]
9. Ettehadi, M.; Ghasemi, H.; Vaez-Zadeh, S. Voltage Stability-Based DG Placement in Distribution Networks. *IEEE Trans. Power Deliv.* **2012**, *28*, 171–178. [CrossRef]
10. De Oliveira, T.E.C.; Bollen, M.H.J.; Ribeiro, P.F.; Carvalho, P.M.; Zambroni, A.C.; Bonatto, B.D.; De Oliveira, T.E.C. The Concept of Dynamic Hosting Capacity for Distributed Energy Resources: Analytics and Practical Considerations. *Energies* **2019**, *12*, 2576. [CrossRef]
11. Błędzińska, M.; Barchi, G.; Paska, A.J. Time-Series PV Hosting Capacity Assessment with Storage Deployment. *Energies* **2020**, *13*, 2524. [CrossRef]
12. Lin, Y.; Ding, T.; Bie, Z.; Li, G. A new method to evaluate maximum capacity of photovoltaic integration considering network topology reconfiguration. In Proceedings of the 2016 IEEE Power and Energy Society General Meeting (PESGM), Boston, MA, USA, 17–21 July 2016; pp. 1–5.
13. Calderaro, V.; Piccolo, A.; Siano, P. Maximizing DG penetration in distribution networks by means of GA based reconfiguration. In Proceedings of the 2005 International Conference on Future Power Systems, Amsterdam, The Netherlands, 18 November 2005; p. 6.

14. Rosseti, G.J.; De Oliveira, E.J.; De Oliveira, L.W.; Silva, I.C.; Peres, W. Optimal allocation of distributed generation with reconfiguration in electric distribution systems. *Electr. Power Syst. Res.* **2013**, *103*, 178–183. [CrossRef]

15. Jakus, D.; Cadenovic, R.; Vasilj, J.; Sarajcev, P. Maximizing distribution network hosting capacity through optimal network reconfiguration. In Proceedings of the 2019 16th International Conference on the European Energy Market (EEM), Ljubljana, Slovenia, 18–20 September 2019; pp. 1–5.

16. Alturki, M.; Khodaei, A. Increasing Distribution Grid Hosting Capacity through Optimal Network Reconfiguration. In Proceedings of the 2018 North American Power Symposium (NAPS), Fargo, ND, USA, 9–11 September 2018; pp. 1–6.

17. Pyone, L.S. Feeder Reconfiguration and Distributed Generator Placement in Electric Power Distribution Network. *Am. J. Electr. Comput. Eng.* **2018**, *2*, 56–63.

18. Celli, G.; Ghiani, E.; Mocci, S.; Pilo, F. A Multiobjective Evolutionary Algorithm for the Sizing and Siting of Distributed Generation. *IEEE Trans. Power Syst.* **2005**, *20*, 750–757. [CrossRef]

19. Ochoa, L.F.; Dent, C.J.; Harrison, G.P. Distribution Network Capacity Assessment: Variable DG and Active Networks. *IEEE Trans. Power Syst.* **2010**, *25*, 87–95. [CrossRef]

20. Rylander, M.; Smith, J.; Sunderman, W. Streamlined Method for Determining Distribution System Hosting Capacity. In Proceedings of the 2015 IEEE Rural Electric Power Conference, Asheville, NC, USA, 24 August 2015; pp. 3–9.

21. Capitanescu, F.; Ochoa, L.F.; Margossian, H.; Hatziargyriou, N.D. Assessing the Potential of Network Reconfiguration to Improve Distributed Generation Hosting Capacity in Active Distribution Systems. *IEEE Trans. Power Syst.* **2014**, *30*, 346–356. [CrossRef]

22. Wang, S.; Chen, S.; Ge, L.; Wu, L. Distributed Generation Hosting Capacity Evaluation for Distribution Systems Considering the Robust Optimal Operation of OLTC and SVC. *IEEE Trans. Sustain. Energy* **2016**, *7*, 1111–1123. [CrossRef]

23. Georgilakis, P.S.; Hatziargyriou, N.D. Optimal Distributed Generation Placement in Power Distribution Networks: Models, Methods, and Future Research. *IEEE Trans. Power Syst.* **2013**, *28*, 3420–3428. [CrossRef]

24. Taylor, J.A.; Hover, F.S. Convex Models of Distribution System Reconfiguration. *IEEE Trans. Power Syst.* **2012**, *27*, 1407–1413. [CrossRef]

25. Wu, W.; Tian, Z.; Zhang, B. An Exact Linearization Method for OLTC of Transformer in Branch Flow Model. *IEEE Trans. Power Syst.* **2017**, *32*, 2475–2476. [CrossRef]

26. Baran, M.; Wu, F. Network reconfiguration in distribution systems for loss reduction and load balancing. *IEEE Trans. Power Deliv.* **1989**, *4*, 1401–1407. [CrossRef]

27. Baringo, L.; Conejo, A. Wind power investment within a market environment. *Appl. Energy* **2011**, *88*, 3239–3247. [CrossRef]

28. General Algebraic Modeling System. GAMS. Available online: https://www.gams.com/ (accessed on 11 August 2020).

29. IBM ILOG CPLEX. IBM. Available online: https://www-01.ibm.com/software/commerce/optimization/cplex-optimizer/ (accessed on 11 August 2020).

Article

Comparison between Inflexible and Flexible Charging of Electric Vehicles—A Study from the Perspective of an Aggregator

Isaias Gomes [1,2], Rui Melicio [1,2,3,*] and Victor Mendes [3,4,5]

1   ICT, Instituto Ciências da Terra, Universidade de Évora, 7002-554 Évora, Portugal; ilrgomes21@gmail.com
2   IDMEC, Instituto Superior Técnico, Universidade de Lisboa, 1049-001 Lisbon, Portugal
3   Departamento de Física, Escola de Ciências e Tecnologia, Universidade de Évora, 7002-554 Évora, Portugal; vfmendes@deea.isel.pt
4   CISE, Electromechatronic Systems Research Centre, Universidade da Beira Interior, 6201-001 Covilhã, Portugal
5   Department of Electrical Engineering and Automation, Instituto Superior de Engenharia de Lisboa, 1959-007 Lisbon, Portugal
*   Correspondence: ruimelicio@gmail.com; Tel.: +35-126-674-5372

Received: 22 July 2020; Accepted: 10 October 2020; Published: 19 October 2020

Abstract: This paper is about the problem of the management of an aggregator of electric vehicles participating in an electricity market environment. The problem consists in the maximization of the expected profit through a formulation given by a stochastic programming problem to consider the uncertainty faced by the aggregator. This uncertainty is due to the day-ahead market prices and the driving requirements of the owners of the vehicles. Depending on the consent of the owners, inflexible charging to flexible charging is considered. Thus, the aggregator can propose different profiles and charging periods to the owners of electric vehicles. Qualitatively, as expected, the more flexible the vehicle owners, the higher the expected profit. The formulation, however, offers more to the aggregator and provides the ability to quantify the influence of consent of favorable driving requirements in the expected profit, allowing the aggregator to consider rewarding the owners of vehicles with more flexibility. Case studies addressed are for comparison of the influence of owners having inflexibility, partial flexibility, or flexibility in the expected profit of the aggregator.

Keywords: battery degradation; electric vehicles; electric vehicles aggregator; electricity markets; stochastic programming

---

1. Introduction

Energy production has been essentially based on non-renewable sources, i.e., from fossil fuels, resulting in increased anthropogenic gas emissions. Among the sectors of the economy, the transport sector has a weight of about 15% in global emissions [1]. Likewise, the transport sector is responsible for more than 28% of the energy consumed in most countries [2]. To reduce greenhouse gas emissions, electric vehicle adoption has been discussed worldwide [3,4]. The market share of electric vehicles remains low since the higher acquisition cost of electric vehicles compared to conventional vehicles is one of the important barriers to adoption [3]. Policy incentives for the acquisition of electric vehicles include financial incentives and non-financial incentives [3]. Financial incentives include direct purchasing subsidies, and registration/emission/tax fee exemptions, which are the most widely used incentives to lower the high initial purchasing cost [3]. Non-financial incentives are designed for the convenience of electric vehicle owners, including free parking policies, and toll tax exemptions [3]. The power system perceives the integration of a large number of electric vehicles as a threat and the electricity market participants perceive this integration as an opportunity to consider electric vehicles as a new source of energy in the electricity market [5]. Therefore, in the coming years, with the increase in electric vehicles in the power system, aggregating agents tend to play an important role as intermediaries among the owners of electric vehicles, the electricity market, distribution system operators, and transmission system operators [6]. Ref. [7] presents one of the most important works regarding the role of electric vehicle aggregators in electricity markets. The trading of energy in electricity markets is a well-studied subject [8–14]. Ref. [8] proposes an aggregation platform for wind, photovoltaic, and thermal power in electricity markets. Ref. [9,12] present the stochastic coordination of joint wind and photovoltaic systems in the day-ahead market. Ref. [10] proposes the coordination of wind power, photovoltaic power, and energy storage. Ref. [11] proposes the analysis of the impact of dust on photovoltaic systems in aggregation with wind and thermal power. Ref. [13] presents the self-scheduling and bidding strategies of thermal units with stochastic emission constraints. Ref. [14] proposes the bidding strategy of wind–thermal energy producers.

In recent years, the participation of electric vehicle aggregators in electricity markets has been gaining attention among researchers and market players. Electric vehicles management has been proposed according to two different approaches: considering unidirectional charging [15], i.e., grid-to-vehicle; and considering bidirectional charging [16], vehicle-to-grid (V2G). The development of management systems for the management of electric vehicles in electricity markets is divided into two main groups: group 1 studies electric vehicles as deterministic units [17–19], suggesting the absence of uncertain characteristics; group 2 studies electric vehicles as stochastic units [17–25], suggesting the consideration of uncertain characteristics. Ref. [17] presents a mechanism to determine the two-way energy storage of a large pool of electric vehicles that can be contracted in the ancillary services market on a long-term basis to provide the regulation up and regulation down to the grid. Ref. [18] proposes an algorithm for an electric vehicle aggregator providing unidirectional V2G regulation. Ref. [20] proposes a bidding strategy for an electric vehicle aggregator that participates in the day-ahead market. The problem is formulated using stochastic robust optimization, considering uncertainty in day-ahead market prices, and driving requirements of electric vehicles. Ref. [21] investigates the application of stochastic dynamic programming to the optimization of charging and frequency regulation capacity bids of an electric vehicle in a smart grid [26,27]. Ref. [22] presents an optimal bidding strategy of an electric vehicle aggregator participating in day-ahead energy and regulation markets using stochastic optimization. Ref. [23] proposes a stochastic optimization model for optimal bidding strategies of electric vehicle aggregators in day-ahead and ancillary services markets with variable wind energy. Ref. [24] proposes the optimal scheduling of plug-in electric vehicle aggregators in the electricity market considering as uncertain parameters market prices, availability of electric vehicles, and status of being called in the reserve market. Ref. [25] proposes a two-stage stochastic optimization problem for optimally coordinated bidding of an electric vehicle in the day-ahead, intra-day and real-time markets. The formulation also includes risk management with the hourly conditional value at risk. Ref. [28] presents the bidding strategy for electric vehicle aggregators participating in the day-ahead market. Ref. [29] proposes the problem of decision making on an electric vehicle aggregator in a competitive market in the presence of different uncertain resources. The problem is formulated using stochastic programming. Ref. [30] proposes a stochastic programming problem for the scheduling of electric vehicle aggregators in the day-ahead market. The main contribution of this paper is the consideration of the level of flexibility of the owners of vehicles. Flexibility is the consent regarding electricity usage for charge/discharge to a practice stated by the aggregator, allowing the aggregator to obtain a higher expected profit than the one in the case of the inflexibility of owners. This contribution allows further improvement in the management of the aggregator and is an extension of the scope of the work in [5] addressing only inflexible owners.

2. Assumptions

This paper considers a set of assumptions: (1) the aggregator is the intermediate entity between the owners of electric vehicles and the electricity market; (2) the parties agree on the inflexibility or flexibility for the electricity usage, so the aggregator is only empowered to carry out operation planning with the owners' authorization; (3) the aggregator is responsible for the degradation of vehicles when in the process of charging or discharging, in which the cost is paid to the owners of electric vehicles; (4) the owners of electric vehicles pay back to the aggregator the cost of the degradation due to the demand for energy for driving; (5) the owners of electric vehicles are penalized with the respective cost of the degradation of the vehicles, when a violation occurs on the schedule given by aggregator for charge or discharge of energy; (6) the market gives aggregators a premium for the participation of electric vehicles in the electricity market, through a special price in the purchase of energy, in order to stimulate the purchase of electric vehicles and their participation in the electricity market and in a wider energy matrix; (7) through an agreement between the parties, the profit obtained from participation in the electricity market can be divided between the aggregator and the owners of electric vehicles.

3. Problem Formulation

The current wholesale markets do not allow small agents to participate in the market, due to minimal power requirements, for example, in Nord Pool, the minimum bid size in the day-ahead market is 1 MW. So, a single electric vehicle is not allowed to participate in the market, but a fleet of electric vehicles satisfying minimal power requirements can be in a market through an aggregator. The electric vehicles are considered in this paper as non-stationary energy storage devices. The aggregator's functions are: (1) the management of charges and discharges in periods where the vehicles are available, by agreement with the owners of the vehicles; (2) to be an intermediary agent between the owners of electric vehicles and the electricity market; (3) present profitable offers for the purchase and sale of energy in the day-ahead market; (4) to manage the charges and discharges to reduce the degradation of the batteries of electric vehicles. The aggregator's main objective in participating in the market is the maximization of profit, subjected to constraints as follows:

$$\max \sum_{s=1}^{N_S} \sum_{t=1}^{N_T} \frac{1}{N_S} \left(\lambda_{st}^{DA} P_{st}^D - \lambda_{st}^{DA*} P_{st}^C + \zeta E_{st}^R - C_{st}^{Deg} \right) \tag{1}$$

Subject to:

$$\underline{P^D} \sigma_{st}^D \leq P_{st}^D \leq \overline{P^D} \sigma_{st}^D \; \forall s, \; \forall t \tag{2}$$

$$\underline{P^C} \sigma_{st}^C \leq P_{st}^C \leq \overline{P^C} \sigma_{st}^C \; \forall s, \; \forall t \tag{3}$$

$$0 \leq \sigma_{st}^D \leq \sigma_{st}^A \; \forall s, \; \forall t \tag{4}$$

$$0 \leq \sigma_{st}^C \leq \sigma_{st}^A \; \forall s, \; \forall t \tag{5}$$

$$\sigma_{st}^D + \sigma_{st}^C \leq \sigma_{st}^A \; \forall s, \; \forall t \tag{6}$$

$$SoC_{st} = SoC_{st-1} + \frac{\eta^C P_{st}^C}{\overline{E}} - \frac{P_{st}^D}{\overline{E} \eta^D} - \frac{E_{st}^R}{\overline{E}} \; \forall s, \; \forall t \tag{7}$$

$$\underline{SoC} \leq SoC_{st} \leq \overline{SoC} \; \forall s, \; \forall t \tag{8}$$

$$P_{st}^D - P_{s't}^D \leq 0 : \lambda_{s't}^{DA} \geq \lambda_{st}^{DA} \; \forall s, s', \; \forall t \tag{9}$$

$$P_{st}^C - P_{s't}^C \leq 0 : \lambda_{st}^{DA} \geq \lambda_{s't}^{DA} \; \forall s, s', \; \forall t \tag{10}$$

$$P_{st}^D = P_{s't}^D : \lambda_{s't}^{DA} = \lambda_{st}^{DA} \; \forall s, s', \; \forall t \tag{11}$$

$$P_{st}^C = P_{s't}^C : \lambda_{s't}^{DA} = \lambda_{st}^{DA} \; \forall s, s', \; \forall t \tag{12}$$

In (1), the objective function is the aggregator's expected profit, with the following terms: (1) revenue from sales offers in the day-ahead market, as a result of electric vehicle discharges, where λ_{st}^{DA} is the day-ahead market price and P_{st}^D is the sale offer/discharge power; (2) cost of purchase offers in the day-ahead market, as a result of charging electric vehicles, where λ_{st}^{DA*} is a V2G tariff to encourage the electric vehicle owners to operate in V2G mode and P_{st}^{C} is the purchase offer/charge power; (3) revenue from the energy consumed by the owners of electric vehicles, where ζ is the price for driving requirements and E_{st}^{R} is the energy consumed for driving requirements; and (4), the cost of battery degradation of electric vehicles given by C_{st}^{Deg}. In (2), the technical limits of operation are presented for the sale offer/discharge power, where P^{D} and $\overline{P^{D}}$ are the minimum and maximum discharge power, respectively. In (3), the technical limits of operating limits are presented for the purchase offer/charge power, where P^{C} and $\overline{P^{C}}$ are the minimum and maximum charge power, respectively. In (2) and (3), σ_{st}^{D} and σ_{st}^{C} are the binary variables that model the discharge and charge cycles of electric vehicles, respectively. In (4) and (5), maximum and minimum values of the binary variables are presented, where σ_{st}^{A} is the availability parameter of electric vehicles (0/1 parameter). In (6), it is imposed that electric vehicle batteries cannot charge and discharge at the same time. According to the fleet being available, there are two events imposed to be feasible and one event imposed to be feasible for the fleet being unavailable: ($\sigma_{st}^{A} = 1$; $\sigma_{st}^{D} = 1$, $\sigma_{st}^{C} = 0$) is event_1, available for discharge; ($\sigma_{st}^{A} = 1$; $\sigma_{st}^{D} = 0$, $\sigma_{st}^{C} = 1$) is event_2, available for charge; ($\sigma_{st}^{A} = 0$; $\sigma_{st}^{D} = 0$, $\sigma_{st}^{C} = 0$) is event_3, in which discharge or charge is unavailable. In (7), the equation of state of charge of the electric vehicle battery equation is presented, where SoC_{st} is the state of charge, η^{C}/η^{D} is the charge/discharge efficiency and $\overline{E}$ is the maximum capacity of the battery of electric vehicles. In (8), $\underline{SoC}$ and $\overline{SoC}$ are the minimum and maximum values of the state of charge variable, respectively. In (9), it is imposed that the offering curves for the sale offer increase monotonically, where $P_{s't}^{D}$ is the power discharge of a specific scenario and $s \neq s'$. In (10), it is imposed that the offering curves for the purchase offer decrease monotonically, where $P_{s't}^{C}$ is the power charge of a specific scenario and $s \neq s'$. Regarding (9) and (10), once the market participants submit their offering curves for sales offers and offering curves for purchase offers, the market operator clears the day-ahead market and publishes the market-clearing price of the day-ahead market and the accepted offers. In (11) and (12), non-anticaptivity constraints are imposed. With these constraints, only one offering curve can be submitted to the day-ahead market for each hour irrespective of the driving requirement realizations. The expression to compute the battery degradation is as follows [31]:

$$ C_{st}^{Deg} = \left| \frac{m}{100} \left(\frac{P_{st}^{C} + P_{st}^{D} - E_{st}^{R}}{\overline{E}} \right) C^{B} \right| \tag{13} $$

In (13), m is the parameter for the relationship of the cost incurred with the reduction in useful life due to charge or discharge of the battery [31] and C^{B} is the cost of batteries for electric vehicles. The extra cost of degradation incurred by charge or discharge processes due to the aggregator management decisions is paid to the electric vehicle owners. This cost is given by the total cost of degradation minus the cost of degradation incurred by the owner driving requirement since the energy used in driving is the one stored by charging.

4. Uncertainty Modeling

4.1. Scenario Generation

The management of electric vehicles in a market environment involves an increased level of uncertainty due to not only market prices that have some characteristics of volatility, but also to the uncertainty on the behavior of owners of the electric vehicles in what regards the usage of electricity. Thus, the uncertain parameters considered in this paper are the day-ahead market prices, the energy demand for electric vehicles for driving requirements, and the availability of electric vehicles. To incorporate the uncertainty into the stochastic model proposed in this paper, the uncertain parameters are modeled through a set of plausible realizations known to be the scenarios. Energy demand for electric vehicles and the availability of vehicles are modeled at coincident periods and scenarios.

The following steps are applied to generate the scenarios: (1) the consideration of the kernel density estimation (KDE), which is a non-parametric method to estimate the probability density function (pdf) of a random variable; (2) from the estimation revealed by KDE, a total of 1000 scenarios are generated. The kernel density estimator is as follows [32]:

$$\hat{f}_h(x) = \frac{1}{nh} \sum_{i=1}^{n} K\left(\frac{x - x_i}{h}\right) \tag{14}$$

In (14) n is the sample size, h is the bandwidth, and $K(.)$ is the kernel smoothing function.

4.2. Scenario Reduction

The 1000 scenarios generated for each parameter in the previous section are reduced to 10 by applying the K-means scenario reduction technique. The K-means technique identifies clusters of scenarios that show considerable similarity; determining a scenario represents clusters of scenarios while maintaining relevant information for optimal decision-making. The K-means technique is relatively easy to implement, having broad convergence, and can be adapted to larger data sets. The objective function of the K-means technique is as follows [33]:

$$\min \sum_{k=1}^{K} \sum_{x_i \in S_k} dist(x_i, c_k)^2 \tag{15}$$

In (15), K is the number of clusters, $dist$ is a chosen distance measure, and x_i and c_k are data points and centroids belonging to cluster S_k.

5. Case Studies

The case studies are tested using data of day-ahead market prices from the Iberian Electricity Market (MIBEL) [34] and the behavior profiles of drivers in Europe [35]. The time horizon considered is 24 h. The electric vehicle aggregator manages 1000 similar electric vehicles, having a battery capacity of 25 kWh [5]—a common value for Nissan Leaf models. The rated power of the super battery/superload is 25 MWh [5]. This paper assumes that the driving patterns of all electric vehicles are similar. The battery cost in an electric vehicle is 250 €/kWh and the battery modeled for battery degradation has the cost given by Equation (13) with a linear approximated slope of m −0.0013 given in [31]. The ratio distance/consumption is assumed to be the same, having a value of 6 km/kWh. The V2G tariff λ_{st}^{DA*} is assumed to be $\lambda_{st}^{DA*} = 0.65\lambda_{st}^{DA}$. This V2G tariff should be sufficiently low to encourage the electric vehicle aggregator to participate in electricity markets, covering the operating costs of the electric vehicles and the battery degradation costs. Similar approaches regarding incentives for aggregators are applied in other research works, such as the case of [36] where a price markup is considered. Three case studies are considered: (1) Case 1—Inflexible charging; (2) Case 2—Partially-flexible charging; (3) Case 3—Flexible Charging.

The analysis of the results obtained from the formulation proposed in this paper is on the basis of the offering curves, whether regarding the offering curves for the purchase or sale of energy, typically presented in electricity markets, as is the case in the Iberian Electricity Market. The purpose of the analysis is to confirm that the flexibility of vehicle owners can influence the management of the electric vehicle fleet and to determine how flexibility can improve the aggregator's profit through the use of the offering curves. The importance of this analysis is its further support for the aggregator to access the market with better levels of rationality when submitting the offers. The scenarios of day-ahead market prices given by the K-means scenario reduction technique are in Figure 1.

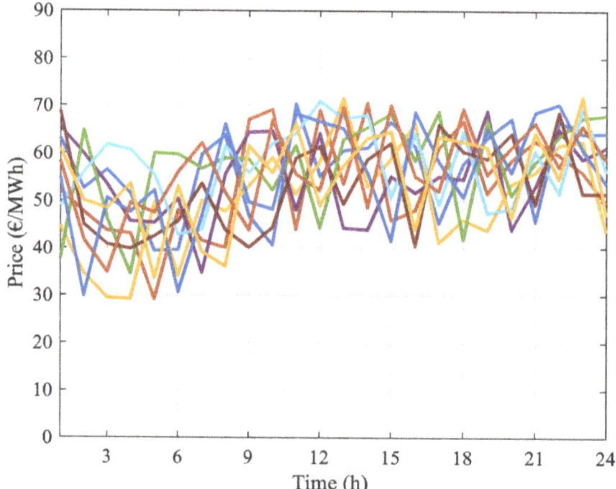

Figure 1. Scenarios of day-ahead market prices.

Figure 1 shows the typical tendency for the behavior of market prices.

5.1. Case_1—Inflexible

In this case study, the electric vehicle aggregator has no power over the behavior of the vehicle, i.e., owners of vehicles are inflexible. The aggregator has power only when electric vehicles are not moving. Thus, through the communication system between the aggregator and the owners, the latter sends the predictable data on their behavior. The owner of the vehicle has the possibility to send 10 scenarios of driving requirements, according to the data in Appendix A, Table A1. The data show that vehicles between hour 1 and hour 7 are parked. In addition, the data show that the vehicles are, not with certainty, parked in periods of likely high day-ahead market prices, namely, hours 10, 13, and 22, and are explicitly driving in hour 21. Consequently, the vehicles are, not with certainty, able to charge or discharge the batteries in those periods. Based on these data, at the time of decision making, the aggregator defines the optimal values and the times for charging and discharging the electric vehicles, respectively, for purchase offers and sales offers, to present optimal offers in the electricity market. So, the aggregator has no opportunity to present offers in periods that have the potential for economic advantage. Purchase offering and sale curves are in Figures 2 and 3, respectively.

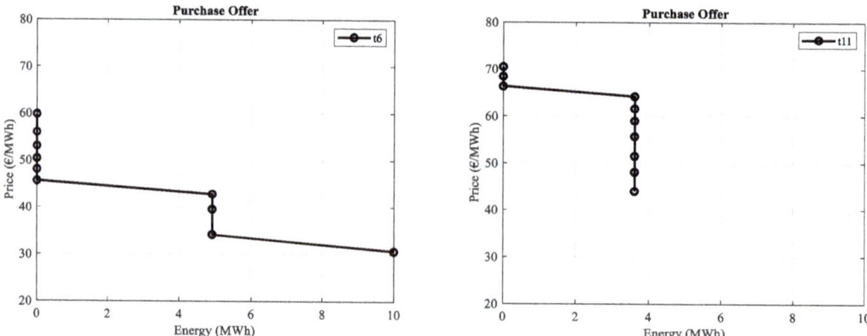

Figure 2. Purchase offering curves: **left**, hour 6; **right**, hour 11.

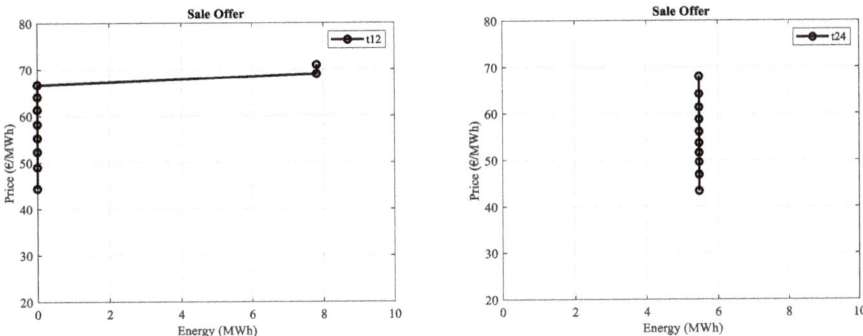

Figure 3. Sale offering curves: **left**, hour 12; **right**, hour 24.

Figure 2 shows the offering curves for purchase offers for hour 6 and hour 11. The offering curves for purchase offers decrease monotonically, as imposed by (10). At hour 6, as the day-ahead market price is one of the lowest, the aggregator takes the opportunity to charge the battery of electric vehicles. However, the aggregator is only willing to buy energy at a price below 45 €/MWh. So, when the price is higher than this value, the offer has a value of 0 MWh. The aggregator is willing to charge the battery of electric vehicles if the purchase price is around 30 €/MWh. At hour 11, as the day-ahead market price is already higher, the aggregator chooses to buy less energy. Thus, the aggregator only buys around 3.5 MWh of energy. To buy this amount of energy, the aggregator is willing to pay a maximum of 65 €/MWh. Above that value, the best option for the aggregator is not to buy energy in the day-ahead market. Figure 3 shows the offering curves for sale offers for hour 12 and hour 24. The offering curve for sale offers increases monotonically, as imposed by (9). At hour 12, one of the periods of the day with the best day-ahead market price, the aggregator is only available to sell energy at a price above 69 €/MWh. Below this value, the offer has a value of 0 MWh. At hour 24, the day-ahead market price is higher than some periods of the time horizon. Then, the aggregator is willing to receive anything between 44 €/MWh and 68 €/MWh for 5.5 MWh of energy. This behavior is the behavior of a perfectly inelastic supply curve. The approach proposed in this paper makes it possible to acquire the offering curves, which allow the aggregator to present bids for offer blocks at the lowest possible price when purchasing energy, and bids for offer blocks at the highest possible price when selling energy. In addition, this approach allows studying strategies to allocate part of the augmented profit among the owners of vehicles, thus allowing for flexibility.

5.2. Case_2—Partially-Flexible

In this case study, the aggregator has power over the behavior of electric vehicle owners in a more extended period. By agreement, the behavior of vehicle owners is less uncertain, and part of the energy is consumed in fewer hours. With the extended period of control over the behavior of vehicles, the aggregator can make better decisions, since periods of better market prices are available to inject energy into the grid and thus improve profit. In this case study, the scenarios of driving requirements and the availabilities of electric vehicles are in accordance with the data in Appendix A, Table A2. The driving hours are just 5 hours per day, giving the aggregator the possibility to implement more profitable decisions without significantly interfering in the routines of vehicle owners. It is worth noting that hours 10, 13, 21, and 22 are flexible hours with high day-ahead market prices. Purchase offering and sale curves are in Figures 4 and 5, respectively.

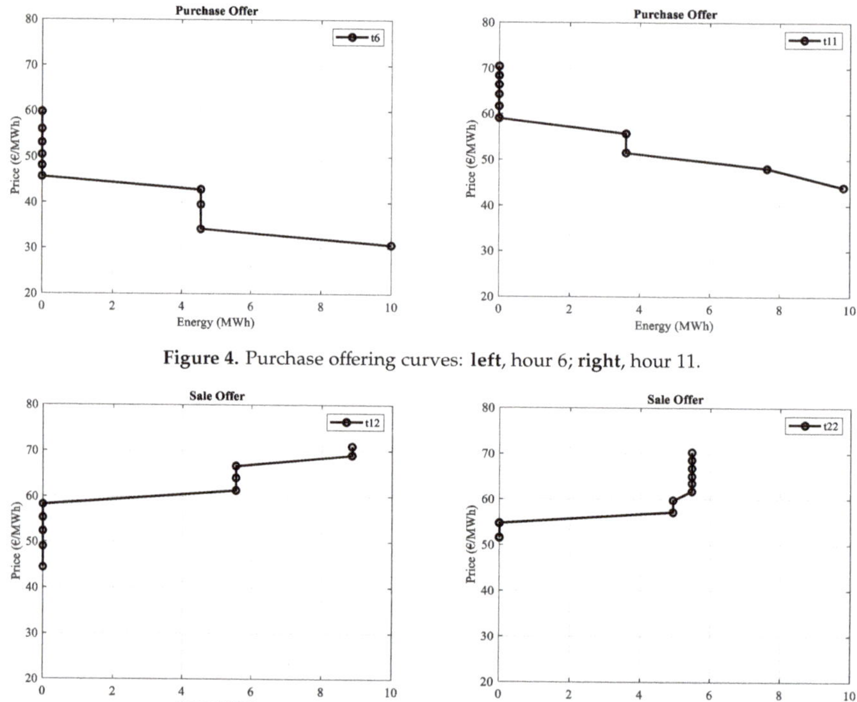

Figure 4. Purchase offering curves: **left**, hour 6; **right**, hour 11.

Figure 5. Sale offering curves: **left**, hour 12; **right**, hour 22.

Figure 4 shows that at hour 6, the aggregator is only available to buy energy below 43 €/MWh. Above this value, the offer is 0 MWh. The aggregator is available to buy 4.5 MWh for values between €34 and €43. To buy 10 MWh, the aggregator only accepts at a value of around €31. The offer values for hour 6 of Case 2 are very similar to the values for hour 6 of Case 1. The difference is in the offer of 4.5 MWh, which in Case 1 is 4.9 MWh. At hour 11, the aggregator is only available to buy energy below 56 €/MWh. Above this value, the offer is 0 MWh. The aggregator is available to buy 3.6 MWh for values between €51 and €56. To buy 7.7 MWh and 9.8 MWh, the aggregator only accepts at a value of around €48 and €44, respectively. Compared to hour 11 in Case 1, the aggregator is more demanding, since the aggregator only accepts to trade below €56, while in Case 1 the trade starts from €64. Figure 5 shows that at hour 12, the aggregator is only available to sell energy above 61 €/MWh. Below this value, the offer is 0 MWh. The aggregator is available to sell 5.5 MWh for values between €61 and €66. To sell 8.8 MWh, the aggregator only accepts at a value of around €69. Compared to hour 12 in Case 1, the aggregator starts to trade from €61, while, in Case 1, the aggregator starts to trade from €69. However, for Case 2, starting at €69, the aggregator only negotiates an amount of energy equal to 8.8 MWh. A convenient consent of the owners of vehicles allows the aggregator, by implementing the strategy that yields a higher profit, to consider rewarding the owners in order to encourage further flexibility.

5.3. Case_3—Flexible

In this case study, the aggregator has full power over the batteries of electric vehicles, and the aggregator decides the charging and discharging periods throughout the day, i.e., by assumption, the owners of the electric vehicle do not use the vehicle during the entire time horizon under study. Purchase offering and sale curves are in Figures 6 and 7, respectively.

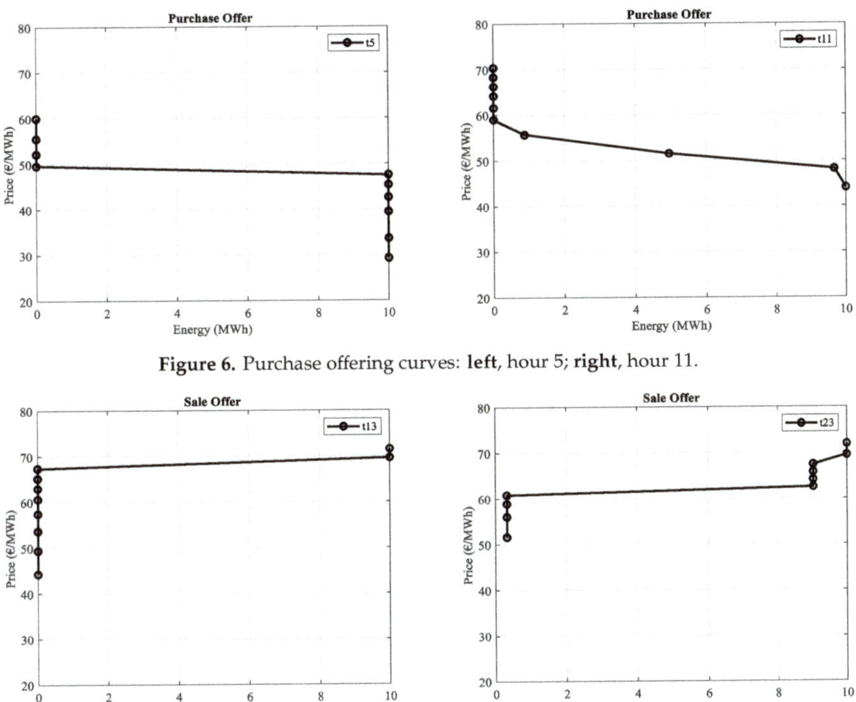

Figure 6. Purchase offering curves: **left**, hour 5; **right**, hour 11.

Figure 7. Sale offering curves: **left**, hour 13; **right**, hour 23.

Figure 6 shows that compared to Case 2, the optimum values suggest that the batteries of electric vehicles be charged at hour 5, an hour with an average market price slightly lower than that of hour 6. Thus, at hour 5, the aggregator is only available to buy energy below 48 €/MWh. Above this value, the offer is 0 MWh. The aggregator is available to buy 10 MWh for values between €29 and €48. Compared to hour 6 in Case 2, the aggregator can buy large quantities of energy at lower prices. At hour 11, the aggregator is only available to buy energy below 56 €/MWh. Above this value, the offer is 0 MWh. Compared to hour 11 in Case 2, the aggregator can buy larger quantities of energy for the same prices. For example, for a price of €44, the aggregator in Case 3 buys 10 MWh, while, in Case 2, the aggregator buys 9.8 MWh, and for a price of €48, the aggregator buys 9.7 MWh in Case 3 and buys 7.6 MWh in Case 2. Figure 7 shows that at hour 13, the aggregator is only available to sell energy above 69.6 €/MWh. Below this value, the offer is 0 MWh. Compared to hour 12 in Case 2, the aggregator can sell larger quantities of energy at better market prices. For example, for a price of €69 at hour 12 of Case 2 and a price of €69.6 for hour 13 of Case 3, the aggregator sells 8.9 MWh and 10 MWh, respectively. Similarly, for a price of €71.5, the aggregator sells 10 MWh, while for €70.9, the aggregator sells only 8.9 MWh at hour 12 of Case 2. Likewise, at hour 23 of Case 3, compared to hour 22 of Case 2, the aggregator can sell larger quantities of energy at better market prices. For example, for a price of €68.6 at hour 12 of Case 2 and a price of €69.5 for hour 23 of Case 3, the aggregator sells 5.47 MWh and 10 MWh, respectively. Similarly, for a price of €71.9, the aggregator sells 10 MWh, while, for €70.3, the aggregator buys only 5.47 MWh at hour 22 of Case 2. The possibility of total control over the batteries allows the aggregator to make better decisions through the knowledge that the aggregator has about the market. The comparison between the cases is in Table 1.

Table 1. Comparison between Case 1, Case 2, and Case 3.

	Expected Profit (€)	Degradation Cost (€)
Case 1	504	254
Case 2	666	333
Case 3	1153	487

Table 1 shows that although the cost of degradation increases, the expected profit in Case 2 and Case 3 is higher than the expected profit in Case 1. The expected profit for Case 2 is 32% higher than the profit for Case 1, and the expected profit for Case 3 is more than 100% higher than the profit for Case 1. Thus, the formulation developed offers aggregator support for the management of the charge/discharge of electricity in the vehicles to improve the expected profit. With full control over electric vehicles, owners of electric vehicles are expected to receive a higher reward for being compliant with flexibility, allowing the aggregator to have a further economic advantage. Note that the V2G tariff is important for an aggregator in electricity markets due to the high costs of battery degradation. Hence, the aggregator can make a profit only with adequate values of V2G tariffs.

6. Conclusions

Although the conventional power system perceives the integration of electric vehicles as only a further new load to be satisfied in due time, these vehicles are new sources of energy and opportunities for business in the electricity market. Nevertheless, the owners of electric vehicles cannot participate in electricity markets due to conditions of minimal power requirements imposed in these markets. Thus, these owners can only through an aggregator achieve sufficient conditions to participate in those markets. The aggregator is the intermediary entity between vehicle owners and electricity markets submitting blocks of offers to buy and sell energy to the market and wanting the achieve the most profitable operation, subject to operational and technical constraints. However, a set of electric vehicles is a fleet of non-stationary energy storage devices not necessarily under the total control of the aggregator. The aggregator is only able to manage charge/discharge of the energy storage devices in periods of consent by the owners of the vehicles. This consent is upon agreement with the owners and, in general, must take into consideration the cost due to the incurred degradation of the energy storage devices due to extra usage. So, a support decision for the fleet of vehicles is a crucial aid to the most favorable management regarding participation in the market, and this paper is a contribution in this regard. The support decision for the fleet of vehicles proposed is a formulation based in a mathematical programming problem written as a maximization of the expected profit in a stochastic programming framework, considering the uncertainty in day-ahead market prices and the driving requirements of electric vehicles.

The consent of owners is the level of flexibility regarding electricity usage for charge/discharge to a practice stated by the aggregator, allowing the aggregator to have some control, which is an opportunity to achieve a higher expected profit than the one in the case of inflexibility. As expected, the assumed level of flexibility accepted by the owners has repercussions in the aggregator management, as shown by a comparison of the addressed case studies assuming that the owners are inflexible, partially inflexible, or flexible.

The application proposed in this paper offers the aggregator support for the management of the charge/discharge of electricity in the vehicles to improve the expected profit. Quantitively identifying the augmented profit is a result of having the flexibility stated by the owners to allow the schedule of charge/discharge of vehicles by the aggregator and allowing for simulating the studies of strategies of persuasion for further consent.

Author Contributions: I.G., R.M. and V.M. conceived the formulation, performed the simulations, analyzed the simulation results, and wrote the paper. All authors have read and agreed to the published version of the manuscript.

Funding: This research received no external funding.

Acknowledgments: This work is funded by Bolsas Camões, IP/Millennium BCP Foundation and funded by: European Union through the European Regional Development Fund, included in the COMPETE 2020 (Operational Program Competitiveness and Internationalization); Foundation for Science and Technology (FCT) under the ICT (Institute of Earth Sciences) project UIDB/04683/2020; Portuguese Funds through the Foundation for Science and Technology (FCT) under the LAETA project UIDB/50022/2020; Portuguese Foundation for Science and Technology (FCT) under the CISE Projects UIDB/04131/2020 and UIDP/04131/2020.

Conflicts of Interest: The authors declare no conflict of interest.

List of Symbols

s	index of scenarios
t	index of periods
λ_{st}^{DA}	day-ahead market price
λ_{st}^{DA*}	V2G tariff to encourage the electric vehicle owners to operate in the vehicle to grid mode
ζ	price for driving requirements
σ_{st}^{D}	binary variable modeling discharge cycles
σ_{st}^{C}	binary variable modeling charge cycles
σ_{st}^{A}	binary parameter for input status of electric vehicles
η^{D}	discharge efficiency of batteries of electric vehicles
η^{C}	charge efficiency of batteries of electric vehicles
E_{st}^{R}	driving requirements of electric vehicles
$\overline{E}$	capacity of battery of electric vehicles
P_{st}^{D}	discharge power of electric vehicles/sale offer
P_{st}^{C}	charge power of electric vehicles/purchase offer
$\underline{P^{D}}/\overline{P^{D}}$	minimum/maximum discharge power
$\underline{P^{C}}/\overline{P^{C}}$	minimum/maximum charge power
SoC_{st}	state of charge of electric vehicles
$\underline{SoC}/\overline{SoC}$	minimum/maximum state of charge
C_{st}^{Deg}	cost of battery degradation
C^{B}	cost of batteries of electric vehicles
m	linear approximated slope of battery life as a function of the number of cycles

Appendix A

The following tables only show the driving requirements at hours of the day that have a line of non-null values; in hours not shown, the values are null for all scenarios.

Table A1. Scenarios of driving requirements (E_{st}^{R}) of the fleet of electric vehicles for Case_1.

Hour\Scenario	1	2	3	4	5	6	7	8	9	10
8	2.083	1.185	1.909	1.340	1.468	1.994	2.165	1.694	1.587	1.808
9	1.339	2.492	1.863	2.221	1.982	1.594	2.373	1.732	2.101	1.472
10	0.000	1.998	1.399	0.000	1.036	1.646	0.000	0.000	1.839	1.922
13	2.661	0.000	1.033	1.859	2.540	0.000	2.139	0.000	2.369	1.486
14	2.482	0.000	0.000	1.715	2.827	2.186	0.000	1.271	0.000	2.708
19	0.000	2.497	1.298	2.330	2.027	0.000	1.723	0.000	0.000	0.000
20	2.723	1.109	1.863	2.263	3.130	2.065	2.927	1.626	1.397	2.495
21	2.321	1.558	1.800	2.221	1.707	2.103	1.628	1.906	1.997	1.490
22	2.992	0.000	2.562	0.000	2.073	1.664	0.000	1.324	2.838	0.000

Table A2. Scenarios of driving requirements (E_{st}^{R}) of the fleet of electric vehicles for Case_2.

Hour\Scenario	1	2	3	4	5	6	7	8	9	10
8	2.083	1.185	1.909	1.340	1.468	1.994	2.165	1.694	1.587	1.808
9	1.339	4.490	3.262	2.221	3.018	3.240	2.373	1.732	3.940	3.393
14	5.143	0.000	1.033	3.574	5.367	2.186	2.139	1.271	2.369	4.194
19	2.321	4.055	3.098	4.551	3.734	2.103	3.351	1.906	1.997	1.490
20	5.715	1.109	4.425	2.263	5.203	3.729	2.927	2.951	4.235	2.495

References

1. Organization for Economic Co-Operation and Development. *OECD: Reducing Transport Greenhouse Emissions: Trends & Data*; International Transport Forum: Leipzig, Germany, 2010.
2. Lane, B.W.; Dumortier, J.; Carley, S.; Siddiki, S.; Clark, K.-S.; Graham, J.D. All plug-in electric vehicles are not the same: Predictors of preference for a plug-in hybrid versus a battery electric vehicle. *Transp. Res. Part D Transp. Environ.* **2018**, *65*, 1–13. [CrossRef]
3. Yao, J.; Xiong, S.; Ma, X. Comparative analysis of national policies for electric vehicle uptake using econometric models. *Energies* **2020**, *13*, 3604. [CrossRef]
4. Chen, L.; Zhang, Y.; Figueiredo, A. Spatio-temporal model for evaluating demand response potential of electric vehicles in power-traffic network. *Energies* **2019**, *12*, 1981. [CrossRef]
5. Gomes, I.L.R.; Melicio, R.; Mendes, V.M.F. Electric vehicles aggregation in market environment: A stochastic grid-to-vehicle and V2G management. In *Technological Innovation for Industry and Service Systems*; Springer: Cham, Switzerland, 2019; pp. 343–352.
6. Lopes, J.P.; Soares, F.J.; Almeida, P.M.R. Integration of electric vehicles in electric power system. *Proc. IEEE* **2011**, *99*, 168–183. [CrossRef]
7. Bessa, R.J.; Matos, M.A. The role of an aggregator agent for EV in the electricity market. In Proceedings of the 7th Mediterranean Conference and Exhibition on Power Generation, Transmission, Distribution and Energy Conversion (MedPower 2010), Agia Napa, Cyprus, 7–10 November 2010; pp. 1–9.
8. Gomes, I.L.R.; Pousinho, H.M.I.; Melicio, R.; Mendes, V.M.F. Aggregation platform for wind-pv-thermal technology in electricity market. In Proceedings of the International Symposium on Power Electronics, Electrical Drives, Automation and Motion, Amalfi, Italy, 20–22 June 2018; pp. 799–804.
9. Gomes, I.L.R.; Pousinho, H.M.I.; Melicio, R.; Mendes, V.M.F. Stochastic coordination of joint wind and photovoltaic systems with energy storage in day-ahead market. *Energy* **2017**, *124*, 310–320. [CrossRef]
10. Gomes, I.L.R.; Melicio, R.; Mendes, V.M.F. Decision making for sustainable aggregation of clean energy in day-ahead market: Uncertainty and risk. *Renew. Energy* **2019**, *133*, 692–720. [CrossRef]
11. Gomes, I.L.R.; Melicio, R.; Mendes, V.M.F. Dust Effect impact on PV in an aggregation with wind and thermal powers. *Sustain. Energy Grids Netw.* **2020**, *22*, 100359. [CrossRef]
12. Gomes, I.L.R.; Pousinho, H.M.I.; Melicio, R.; Mendes, V.M.F. Bidding and optimization strategies for wind power-pv systems in electricity markets assisted by CPS. *Energy Procedia* **2016**, *106*, 111–121. [CrossRef]
13. Laia, R.; Pousinho, H.M.I.; Melicio, R.; Mendes, V.M.F. Self-scheduling and bidding strategies of thermal units with stochastic emission constraints. *Energy Convers. Manag.* **2015**, *89*, 975–984. [CrossRef]
14. Laia, R.; Pousinho, H.M.I.; Melicio, R.; Mendes, V.M.F. Bidding strategy of wind-thermal energy producers. *Renew. Energy* **2016**, *99*, 673–681. [CrossRef]
15. Vandael, S.; Claessens, B.; Hommelberg, M.; Holvoet, T.; Deconinck, G. A scalable three-step approach for demand side management of plug-in hybrid vehicles. *IEEE Trans. Smart Grid* **2013**, *4*, 720–728. [CrossRef]
16. Vaya, P.; Baringo, L.; Krause, T.; Andersson, G.; Almeida, P.; Geth, F.; Rapoport, S. EV aggregation models for different charging scenarios. In Proceedings of the 23rd International Conference on Electricity, Lyon, France, 15–18 June 2015; pp. 1–5.
17. Jain, P.; Das, A.; Jain, T. Aggregated electric vehicle resource modeling for regulation services commitment in power grid. *Sustain. Cities Soc.* **2019**, *45*, 439–450. [CrossRef]
18. Sortome, E.; Sharkavi, M.A.-E. Optimal charging strategies for unidirectional V2G. *IEEE Trans. Smart Grid* **2011**, *2*, 131–138. [CrossRef]

19. Naharudinsyah, I.; Limmer, S. Optimal charging of electric vehicles with trading on the intraday electricity market. *Energies* **2018**, *11*, 1416. [CrossRef]

20. Baringo, L.; Amaro, R.S. A stochastic robust optimization approach for the bidding strategy of an electric vehicle aggregator. *Electr. Power Syst. Res.* **2017**, *146*, 362–370. [CrossRef]

21. Donadee, J.; Ilic, M.D. Stochastic optimization of grid to vehicle frequency regulation capacity bids. *IEEE Trans. Smart Grid* **2014**, *5*, 1061–1069. [CrossRef]

22. Vagropoulos, S.I.; Bakirtzis, A.G. Optimal bidding strategy for electric vehicle aggregators in electricity markets. *IEEE Trans. Power Syst.* **2013**, *28*, 4031–4041. [CrossRef]

23. Wu, H.; Shahidehpour, M.; Alabdulwahab, A.; Absusorrach, A. A game theoric approach to risk-based optimal bidding strategies for electric vehicle aggregators in electricity markets with variable wind energy resources. *IEEE Trans. Sustain. Energy* **2016**, *7*, 374–385. [CrossRef]

24. Manijeh, A.; Mohammadi, B.-I.; Moradi, M.-D.; Zare, K. Stochastic scheduling of aggregators of plug-in electric vehicles for participation in energy and ancillary service markets. *Energy* **2017**, *118*, 1168–1179.

25. Vardanyan, Y.; Madsen, H. Optimal coordinated bidding of a profit maximizing, risk-averse EV aggregator in three-settlement markets under uncertainty. *Energies* **2019**, *12*, 1755. [CrossRef]

26. Batista, N.C.; Melicio, R.; Matias, J.C.O.; Catalão, J.P.S. ZigBee standard in the creation of wireless networks for advanced metering infrastructures. In Proceedings of the 16th IEEE Mediterranean Electrotechnical Conference (MELECON'2012), Medina Yasmine Hammamet, Tunisia, 10 May 2012; pp. 220–223.

27. Batista, N.C.; Melicio, R.; Mendes, V.M.F. Services enabler architecture for smart grid and smart living services providers under industry 4.0. *Energy Build.* **2017**, *141*, 16–27. [CrossRef]

28. Guo, Y.; Liu, W.; Wen, F.; Salam, A.; Mao, J.; Li, L. Bidding strategy for aggregators of electric vehicles in day-ahead electricity markets. *Energies* **2017**, *10*, 144. [CrossRef]

29. Rashidizadeh-Kermani, H.; Najafi, H.R.; Anvari-Moghaddam, A.; Guerrero, J.M. Optimal decision-making strategy of an electric vehicle in short-term electricity markets. *Energies* **2018**, *11*, 2413. [CrossRef]

30. Aliasghari, P.; Mohammadi-Ivatloo, B.; Abapour, M.; Ahmadian, A.; Elkamel, A. Goal programming application for contract pricing of electric vehicle aggregator in join day-ahead market. *Energies* **2020**, *13*, 1771. [CrossRef]

31. Sarker, M.R.; Dvorkin, Y.; Ortega, M.A.-V. Optimal participation of an electric vehicle aggregator in day-ahead energy and reserve markets. *IEEE Trans. Syst.* **2016**, *31*, 3506–3515. [CrossRef]

32. Martinez, W.L.; Martinez, A.R. *Computational Statistics Handbook with MATLAB*, 3rd ed.; CRC Press: Boca Raton, FL, USA, 2016.

33. Viegas, J.L.; Vieira, S.M.; Melicio, R.; Mendes, V.M.F.; Sousa, J.M.C. Classification of new electricity costumers based on surveys and smart metering data. *Energy* **2016**, *107*, 804–817. [CrossRef]

34. REE-Red Elétrica de España. Available online: https://www.esios.ree.es/es (accessed on 9 March 2020).

35. Pasaoglu, G.; Fiorello, D.; Martino, A.; Scarcella, G.; Alemanno, A.; Zubaryeva, A.; Thiel, C. *Driving and Parking Patterns of European Car Drivers—A Mobility Survey*; European Commission Joint Research Centre: Luxembourg, 2012.

36. Ortega-Vazquez, M.A.; Bouffard, F.; Silva, V. Electric Vehicle Aggregator/System Operator Coordination for Charging Scheduling and Services Procurement. *IEEE Trans. Power Syst.* **2013**, *28*, 1806–1815. [CrossRef]

Publisher's Note: MDPI stays neutral with regard to jurisdictional claims in published maps and institutional affiliations.

Article

Combined Economic Emission Dispatch with Environment-Based Demand Response Using WU-ABC Algorithm

Ho-Sung Ryu and Mun-Kyeom Kim *

School of Energy System Engineering, Chung-Ang University, 84 Heukseok-ro, Dongjak-gu, Seoul 06974, Korea; ghtjd0580@cau.ac.kr
* Correspondence: mkim@cau.ac.kr; Tel.: +82-2-5271-5867

Received: 4 November 2020; Accepted: 4 December 2020; Published: 6 December 2020

Abstract: Owing to the growing interest in environmental problems worldwide, it is essential to schedule power generation considering the effects of pollutants. To address this, we propose an optimal approach that solves the combined economic emission dispatch (CEED) with maximum emission constraints by considering demand response (DR) program. The CEED consists of the sum of operation costs for each generator and the pollutant emissions. An environment-based demand response (EBDR) program is used to implement pollutant emission reduction and facilitate economic improvement. Through the weighting update artificial bee colony (WU-ABC) algorithm, the penalty factor that determines the weighting of the two objective functions is adjusted, and an optimal operation solution for a microgrid (MG) is then determined to resolve the CEED problem. The effectiveness and applicability of the proposed approach are demonstrated via comparative analyses at a modified grid-connected MG test system. The results confirm that the proposed approach not only satisfies emission constraints but also ensures an economically superior performance compared to other approaches. These results present a useful solution for microgrid operators considered environment issues.

Keywords: combined economic emission dispatch; environment-based demand response; emission constraints; penalty factor; weighting update artificial bee colony

---

1. Introduction

Economic dispatch (ED) is one of the most important issues pertaining to the operation and control of power systems [1,2]. It involves finding an optimal scheduling for all committed generators to minimize fuel costs, while satisfying various constraints such as load demand balance and generation capacity constraints. However, owing to increasing global concerns regarding the environmental issues caused by the combustion of fossil fuels, ED has become a significant concern from an economical perspective and for dealing with pollutant emissions from fossil fuel combustion [3]. Some countries have set maximum emission limits and impose fines when these limits are exceeded [4]. Thus, emission constraints should also be considered during operational scheduling. Economic emission dispatch (EED) has been proposed for scheduling to account for the emission of harmful gases such as carbon dioxide (CO_2), sulfur oxides (SO_x), and nitrogen oxides (NO_x), which pollute the air and exacerbate global warming [5]. The objective function of the EED usually adds emission criteria to the fuel cost of the thermal unit. To integrate the emission component with ED, Dhillon et al. [6] and Kulkarni et al. [7] presented a method whereby a penalty factor is multiplied to the emission term in the objective function. This technique allows both components to become commensurately involved in the optimization.

Demand response (DR) is expected to help solve environmental problems, as it is environmentally friendly and offers several benefits to the entire system [8,9]. This is because DR resources are one of the most inexpensive resources and can react quickly to the commands of the system operator [10]. Furthermore, they do not emit pollutants and help reduce peak loads. In a DR program, consumers sign a contract with the local utility to reduce their demand when requested by the system operator. The utility possesses the advantage of reducing the maximum demand and consequently decreasing operation costs. Moreover, the system operator can use the DR program to satisfy constraints such as maximum pollutant emissions. DR programs are commonly used in small power systems such as microgrids (MGs), because they are difficult to implement in large power systems due to communication problems [11,12].

MGs can be defined as a small-scale form of the centralized power system and typically consist of diesel generation (DG) units, renewable energy resources, and loads that are designed and situated close to customers in small communities [13]. MG operators (MGOs) manage a cluster of loads and power resources, conducting operations to control power locally. MGOs also create contracts with the utility service to meet environmental constraints of the EED, and they take the maximum emission constraints into account in the context of MG operation. To satisfy such constraints, ecofriendly energy resources such as DR should be used appropriately.

Various approaches have been recently adopted to address environmental problems in MG operation. In [14], emissions of harmful gases such as CO_2, SO_x, and NO_x were reduced by installing additional ecofriendly generators that consume less fuel. In [15], a differential evolution technique that combines heat and power was proposed to solve the microgrid economic and emission dispatch problem. In [16], both emission and fuel costs were implemented through different variants of particle swarm optimization. A hybrid algorithm that combines the differential evolution algorithm and particle cluster optimization was used to solve the EED problem [17]. In [18], a price penalty factor, which is the ratio between the maximum fuel cost and the maximum emission of the corresponding generator, was applied for solving the EED problem. However, the studies mentioned above did not take the DR program into account, and the optimal operations were insufficient under environmental considerations. In [19], a multi-objective approach was proposed to optimize microgrid in a short-term with renewable energy sources with a randomized natural behavior. However, pollutant emissions was not considered. Through the implementation of the DR program, in [20], a balanced solution was obtained where a sample hub energy system containing conventional and renewable energy sources was solved with a mixed integer linear program for optimizing operation costs and reducing pollutant emission. In [21], extended game theory was incorporated into the multiobjective dynamic EED optimization problem through a DR model. In [22,23], DR was considered for spinning reserves to solve the EED problem. However, the cost of interruptions and the incentives of the participating subjects were not considered.

Several methods have been considered to solve the EED problem. A flower pollination algorithm was used to solve economic load dispatch and EED problems by considering the power limits of the generator [24]. In [25], economic load and emission dispatch were determined with a bioinspired algorithm in a renewable integrated islanded MG. A converted single objective function comprising fuel cost and emission was optimized using a gravitational search algorithm in [26]. In [27], quantification of the impacts of the CO2 emission has been considered on the planning, economic and scheduling decisions. While most EED problems include pollutant emissions in the objective function, there were no maximum emission constraints on the power system. To operate MGs while accounting for environmental concerns, because maximum emission constraints are practically essential, the optimal operation approach for these problems must be studied. In addition, DR is a considerably ecofriendly energy resource; however, very few studies focus on it to determine volume with respect to environmental constraints. Therefore, it is essential to conduct a study on the optimization of MG operation by utilizing a DR program and considering emission constraints for combined economic emission dispatch (CEED) problems.

In this study, we propose an approach to solve the CEED problem while considering maximum emission constraints. The CEED function consists of the operation costs and pollutant emissions.

Here, the penalty factor is used to convert the emission into an equivalent cost value. Various constraints are considered in solving the problem. In particular, the maximum emission constraint is additionally considered. To solve this problem optimally, we propose an environment-based demand response (EBDR) program that determines volume by optimizing economics and reducing the degree of pollutant emissions from the system. A weighting update artificial bee colony algorithm (WU-ABC) is applied to solve the CEED problem under a maximum emission constraint. Here, the penalty factor is updated using the weighting update method until all the constraints are satisfied. By using WU-ABC algorithm, the proposed approach becomes capable of finding the optimal CEED solution of the MG.

The primary contributions of this paper can be summarized as follows:

- The CEED problem, with a maximum emission constraint on the MG, involves the minimization of operation costs and pollutant emissions. A penalty factor is used by the MGO to control the importance of both objective functions with respect to pollutant emissions. This approach is sufficiently effective to satisfy the maximum emission constraints on MG operation.
- EBDR is applied and both economic and environmental factors are considered. Through the concept of elasticity, the MGO can command demand shifts and reduction, and participants in the DR program are provided appropriate incentives. Through this EBDR program, economics can be improved and environmental hazards can be reduced.
- To solve the CEED problem, we propose the use of the WU-ABC algorithm. This algorithm can efficiently update the penalty factor according to the maximum emission constraint by using the weighting update method in combination with the ABC algorithm, which has been widely used for the optimal operation of power systems of late.

The proposed approach provides an optimal power scheduling using EBDR for MG systems in CEED problems with maximum emission constraints. In this regard, MGOs can be provided with more reasonable and flexible solutions for optimal operation with environmental constraints.

The remainder of this paper is organized as follows. Section 2 introduces the grid-connected MG system model. Section 3 presents the formulation of the problem for the CEED with a maximum emission constraint. Section 4 outlines the proposed EBDR strategy. Section 5 presents the solution of the WU-ABC algorithm and summarizes the overall process. Section 6 presents the simulation results, and Section 7 presents the conclusions of this study.

2. Grid-Connected MG System

Figure 1 shows the structure of the proposed grid-connected MG that includes DG, photovoltaic (PV), wind turbine (WT) units and controllable demands; it is connected to the main grid. MGOs can sell surplus power or buy insufficient power through interactions with the main grid, and they can request demand reduction from the customers using the DR program [28,29].

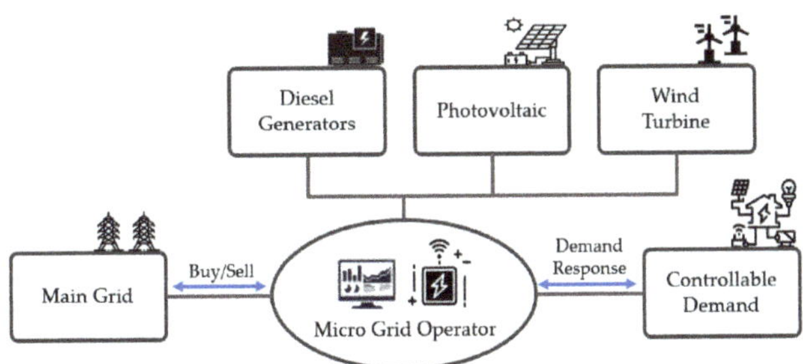

Figure 1. Proposed grid-connected microgrid (MG) system.

2.1. DG Model

The power generation costs of a diesel generator are generally expressed as the sum of a quadratic and sinusoidal function.

$$C_{G,i}(t) = a_i P_{G,i}^2(t) + b_i P_{G,i}(t) + c_i + \left| d_i \sin[P_{G,i}^{min}(t) - P_{G,i}(t)] \right| \tag{1}$$

Here, the sinusoidal function represents a valve point effect to practically account for the generation cost function. Figure 2 illustrates the valve point effect for a conventional generator. It represents a sharp increase in losses due to the wire drawing effect caused by the opening of each steam admission valve [30].

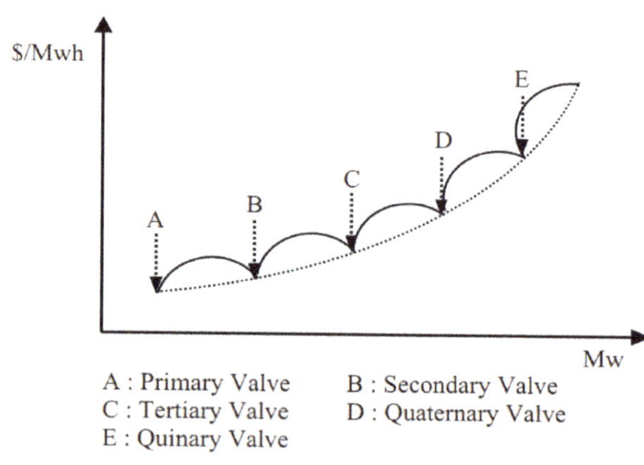

Figure 2. Valve point effect.

2.2. PV Model

The output power of a PV is dependent on solar irradiation. Forecasted solar irradiation is commonly determined using the beta probability distribution function (PDF) expressed as follows [31].

$$PDF_B(si) = \begin{cases} \dfrac{\Gamma(a+b)}{\Gamma(a)\Gamma(b)} \cdot si^{(a-1)} \cdot (1-si)^{(b-1)} & 0 \le si \le 1,\, a \ge 0,\, b \ge 0 \\ 0 & otherwise \end{cases} \tag{2}$$

$$a = \frac{\mu_s \times b}{1 - \mu_s} \tag{3}$$

$$b = (1 - \mu_s) \times \left(\frac{\mu_s \times (1 + \mu_s)}{\sigma_s^2} - 1 \right) \tag{4}$$

The shape parameters a and b are determined according to the mean (μ_s) and standard deviation (σ_s) of solar irradiation data.

Mathematically, the generated power from the PV array is represented as

$$P_s(t) = \eta_s \times A_s \times SI(1 + \beta(T_t - 25)) \tag{5}$$

The generated solar power is represented by the product of PV panel efficiency, size of the PV array, and solar irradiation. β denotes temperature coefficients of the maximum power of the PV array.

2.3. WT Model

The Weibull PDF has been regularly used to model wind speed and can be expressed as

$$PDF_w(v) = \left(\frac{d}{C}\right)\left(\frac{v}{C}\right)^{d-1} \exp\left[-\left(\frac{v}{C}\right)^d\right] \tag{6}$$

The hourly output of a WT unit is highly dependent on wind speed. The mathematical expression for converting the hourly wind speed to power is as follows:

$$P_w(t) = \begin{cases} 0 & if\ v_t < v_{ci} \\ P_r \times \frac{(v_t - v_{ci})}{(v_r - v_{ci})} & if\ v_{ci} \le v_t < v_r \\ P_r & if\ v_r \le v_t < v_{co} \\ 0 & if\ v_{co} \le v_t \end{cases} \tag{7}$$

2.4. Utility Model

For MG operation, MGOs purchase deficit energy from the main grid depending on price and internal generation levels. Conversely, they sell excess energy to the main grid. The amount of power and exchange cost traded between the utility and the MG in each hour is expressed as:

$$P_u(t) = P_{buy}(t)\, b_u(t) - P_{sell}\, (b_u(t) - 1)) \tag{8}$$

$$C_u(t) = \begin{cases} \lambda_{sell}(t) \times P_u(t) & if\ P_u(t) < 0 \\ 0 & if\ P_u(t) = 0 \\ \lambda_{buy}(t) \times P_u(t) & if\ P_u(t) > 0 \end{cases} \tag{9}$$

Here, the power generation state of the MG $b_u(t)$ takes the value 1 when there is a deficit in power, and 0 when there is a surplus of power.

3. Problem Formulation

In the optimization for the CEED problem, two competing objective functions are mathematically formulated by nonlinear functions, which minimize both generation costs and emission function by fulfilling the equality and inequality constraints. In other words, the problem is formulated for the multiobjective function as the minimization of the operation costs and pollutant emissions. The emission function is converted to cost by a multiplying with a weight called the penalty factor.

3.1. Objective Function

3.1.1. Generation Cost Function

The generation cost function is the sum of costs for each generator, including renewable energy source (RES). Conventional generators such as thermal generating units are generally expressed as the sum of a quadratic and sinusoidal function. Costs for PV and WT units, such as the available RES, are determined by considering investment and maintenance costs. The operation costs can be formulated as:

$$F(P) = \sum_{t=1}^{24}[\sum_{i=1}^{I} C_{G,i}(t) + \sum_{j=1}^{J} C_{s,j}P_{s,j}(t) + \sum_{k=1}^{K} C_{w,k}P_{w,k}(t) + C_u(t)] \tag{10}$$

$$P = [P_{G,1}, \ldots, P_{G,I}, P_{s,1}, \ldots, P_{s,J}, P_{w,1}, \ldots, P_{w,K}]^T \tag{11}$$

$$C_{s,j} = AC_{s,j}I_{s,j} + G_{s,j} \tag{12}$$

$$C_{w,k} = AC_{w,k}I_{w,k} + G_{w,k} \tag{13}$$

Equation (11) denotes the vector of the real power output of each generator. Equations (12) and (13) represent the costs of power generation with renewable energy sources, which include installation and maintenance costs. AC is an annuitization coefficient, and it can be calculated using the following equation [32]:

$$AC = \frac{r}{1 - (1+r)^{-N}} \tag{14}$$

3.1.2. Emission Function

The quantity of atmospheric pollutants, such as CO_2, SO_x, and NO_x emitted by a conventional generator can be expressed as the sum of a quadratic and exponential function:

$$E(P) = \sum_{t=1}^{24} em(t) = \sum_{t=1}^{24} \sum_{i=1}^{I} [\{\alpha_i P_{G,i}^2(t) + \beta_i P_{G,i}(t) + \gamma_i + \zeta_i \exp(\lambda_i P_{G,i}(t))\}] \tag{15}$$

In Equation (15), the emission dispatch function is considered as a convex polynomial, similar to the operation cost function. In addition, it is assumed that RES is not considered in the emission function because it does not release air pollutants [33].

3.1.3. Combined Economic Emission Dispatch

As discussed above, the generation costs and emission dispatch function are two different objectives. Hence, a compromise solution is required to solve the CEED problem that minimizes generation costs and the emitted quantities of pollutants. To solve a multiobjective function, a penalty factor is used to reform the emission criteria into the generation cost. Mathematically, the penalty factor is a multiplied value that transforms two different functions into a single objective function. In our work, a DR program is considered for microgrid operation. Therefore, the costs of participating in the DR program are included in the CEED problem, as follows.

$$C(P) = \min \sum_{t=1}^{24} \sum_{i=1}^{I} [F(P) + h_i \times E(P) + C_{DR}(t)] \tag{16}$$

$$C_{DR}(t) = I(\Delta D(t)) + P(\Delta D(t)) \tag{17}$$

Here, the penalty factor h_i indicates the ratio (\$/kg) of the generation costs and emission quantity of each generator i. This factor is initially determined by considering the minimum and maximum of each objective function, and then it is updated to account for environmental constraints. Equation (13) indicates the costs of participating in the DR program; it includes incentive and penalty costs. In this paper, it is determined based on the proposed EBDR program.

3.2. Constraints

3.2.1. Generation Capacity Constraint

The power output of each generation unit is restricted by lower and upper limits for stable operation:

$$P_{G,i}^{\min}(t) \leq P_{G,i}(t) \leq P_{G,i}^{\max}(t) \tag{18}$$

$$P_u^{\min}(t) \leq P_u(t) \leq P_u^{\max}(t) \tag{19}$$

$$P_{s,j}^{\min}(t) \leq P_{s,j} \leq P_{s,j}^{\max}(t) \tag{20}$$

$$P_{w,k}^{\min}(t) \leq P_{w,k}(t) \leq P_{w,k}^{\max}(t) \tag{21}$$

Equations (18)–(21) indicate the minimum and maximum power limits of DG, utility, PV, and WT, respectively.

3.2.2. Power Balance Constraint

The total power generation must cover the total power demand in the presence of transmission line loss:

$$\sum_{i=1}^{I} P_{G,i}(t) + \sum_{j=1}^{J} P_{PV,j}(t) + \sum_{k=1}^{K} P_{WT,k}(t) = P_D(t) - P_{loss}(t) \tag{22}$$

$$P_D(t) = D_0(t) + \Delta D(t) \tag{23}$$

Equation (23) represents the power demand that is finally determined through the DR.

3.2.3. DR Capacity Constraints

The DR capacity is determined by the contract between the MGO and the customers, and it cannot exceed the maximum capacity:

$$\Delta D(t) \leq \Delta D^{\max}(t) \tag{24}$$

3.2.4. Transmission Line Constraints

For stable operation, MGO should consider the transmission line constraint. The upper limit of the transmission line is as follows:

$$S_l \leq S_l^{\max}, \quad l = 1, \ldots, L \tag{25}$$

3.2.5. Emission Constraints

The proposed CEED problem takes into account maximum emission constraints to prevent air pollution. If the MGOs violate emission constraints, they will have to pay a fee for the volume of excess.

$$em(t) \leq em^{\max}(t) \tag{26}$$

4. Environment-Based Demand Response Program

This paper proposes a new type of incentive DR program called EBDR with emission control measures by including emission constraints. Generally, aiming to reduce peak demand, the MGO uses DR programs to adjust consumer behavior and manage power demand. The DR program includes the modification of electricity consumption patterns and incentives to promote this change. Demand and pollutant emission are directly related during MG operation. Thus, when solving the CEED problem, the EBDR is used to control pollutant emissions as well as reduce peak demand for economic benefits. By utilizing EBDR, MGOs can operate the MGs in an ecofriendly manner.

4.1. Price Elasticity of Demand

The EBDR includes the concept of elasticity based on market price. Elasticity is a measure used in economics to assess the percentage of change in demand caused by price fluctuations. In terms of power consumption, this percentage changes as the power demand varies with the changes in market prices, which is defined as an increase in price over time and not an absolute value. Elasticity is expected to be negative because higher power prices can cause demand reductions. The elasticity of the demand for electricity is calculated as follows:

$$E(t_1, t_2) = \frac{MP_0(t_2)}{D_0(t_1)} \times \frac{\partial D(t_1)}{\partial MP(t_2)}, \quad t_1, t_2 = 1, 2 \ldots, 24 \tag{27}$$

Two types of elasticity of demand exist: self-price elasticity and cross-elasticity [34,35]. Self-price elasticity disregards the shift in demand from one period to another and considers changes in consumption over a given period. In this case, increments in market price lead to demand reduction, and self-elasticity always has a negative value. Meanwhile, cross-elasticity is the transferred demand from the peak period to the off-peak demand period within a day. When t_1 is not equal to t_2, the price drop at t_2 causes a demand reduction at t_1. Thus, cross-elasticity always has a positive value.

$$\begin{cases} E(t_1, t_2) \leq 0 & if \ t_1 = t_2 \\ E(t_1, t_2) > 0 & if \ t_1 \neq t_2 \end{cases} \tag{28}$$

4.2. Modeling of EBDR

The MGO enters into contracts with the customers through the EBDR program and mentions the values of incentive and penalty. The customers change their demand from $D_0(t)$ to $D(t)$ according to the contract. The values of incentive and penalty for DR are given by

$$\Delta D(t) = D(t) - D_0(t) \tag{29}$$

$$I(\Delta D(t)) = inc(t) \cdot [D_0(t) - D(t)] \tag{30}$$

$$P(\Delta D(t)) = pen(t) \cdot \{CP(t) - [D_0(t) - D(t)]\} \tag{31}$$

Equation (29) represents the shift in demand of a customer due to the contract. Equations (30) and (31) denote the total values of incentive and penalty, respectively.

The EBDR program shifts the peak demand to reduce operation costs using cross-elasticity based on market prices. The relationship between market prices and demand is represented as follows:

$$D(t_1) = D_0(t_1) + \sum_{t_1=1}^{24} E(t_1, t_2) \cdot \frac{D_0(t_1)}{MP_0(t_2)} \cdot [MP(t_2) - MP_0(t_2)] \quad if \ t_1 \neq t_2 \tag{32}$$

Equation (32) represents the power demand at t_1 according to the market price at t_2, considering a 24-h interval. When incentive and penalty are included in the price, the formula is modified as follows:

$$D(t_1) = D_0(t_1) \left\{ 1 + \sum_{\substack{t_1=1 \\ t_1 \neq t_2}}^{24} E(t_1, t_2) \cdot \frac{[MP(t_2) - MP_0(t_2) + inc(t_2) + pen(t_2)]}{MP_0(t_2)} \right\} \tag{33}$$

Equation (33) indicates only the change in demand due to the market price, without considering emission constraints. The change in demand when emission constraints are included is calculated by utilizing the concept of self-elasticity as follows:

$$\begin{cases} D_e(t_1) = D(t_1) + E_e(t_1, t_1) \cdot \frac{D(t_1)}{em_{max}(t)} \cdot [em(t_1) - em_{max}(t)] & if \ em(t_1) > em_{max}(t_1) \\ D_e(t_1) = D(t_1) & if \ em(t_1) \leq em_{max}(t_1) \end{cases} \tag{34}$$

Here, emission constraints are imposed when emissions are greater than the set emission limit. The above equation yields the 24-h interval consumption of customers participating in the EBDR program, in order to minimize MG operating costs while considering emission constraints.

5. Solution Method

WU-ABC Optimization

In this study, we propose a solution through the WU-ABC algorithm, a hybrid of the weighting update method and the ABC algorithm, to solve the CEED optimization problem. The ABC algorithm is a swarm-based optimization method that simulates the foraging behavior of a bee swarm; it has recently been used in many studies to solve optimization problems [36,37]. However, it is difficult to solve the proposed CEED problem under emission constraints using the ABC algorithm alone. Therefore, the weighting update method is also employed to solve the emission problem by adjusting the penalty factor and EBDR capacity. The WU-ABC algorithm involves five main steps, as described below:

- *Initialization step*: In this step, the initial population of food source is placed randomly in a D-dimensional problem space.

$$x_{mn} = x_{mn}^{\min} + rand[0, 1] \cdot (x_{mn}^{\max} - x_{mn}^{\min}) \tag{35}$$

Equation (35) represents the mth random food source of dimension n in the CEED problem.
- *Employed bees step*: Each bee repeatedly explores a food source to find the optimal solution, and then it chooses a new optimal position (v_{mn}) close to the reference position.

$$v_{mn} = x_{mn} + \phi_{mn} \cdot (x_{mn} - x_{on}), \quad m \neq o \tag{36}$$

When an employed bee finds a food source that is better than the reference one (x_{mn}), the reference food source is superseded with the new candidate.
- *Onlooker bees step*: Onlooker bees look for new positions that are close to the old position through a greedy search method. The bees consider a fitness value, the amount of nectar, and probabilistically determine the space for their next exploration, as follows:

$$P_m = \frac{fit_m}{\sum\limits_{m=1}^{SN} fit_m} \tag{37}$$

In Equation (37), fit_m is a value that is proportional to the amount of nectar.
- *Scout bee step*: If the reference solution cannot be improved through the number of predetermined trials, this food source is abandoned, and the scout bee tries to find a new food source to replace the abandoned food source. The new food source in the solution space is determined using Equation (35). This process is repeated for maximum cycles.
- *Weighting update step*: This step is executed when an optimal solution violates the emission constraint. The penalty factor is updated as much as the actual pollutant emission exceeds the maximum emission, through a weight adjustment method. This process is repeated until the pollutant emission constraints are satisfied for optimal MG operation. Multiplication by a linear function is generally used as the weight update rule. In our work, to improve the convergence speed and accuracy, we use multiplication with an exponential function for the weighting update, as follows:

$$h(s+1) = h(s) \times \exp\left[\eta \times \left(\frac{em(s) - em^{\max}(s)}{em^{\max}(s)}\right)\right] \quad if \ em(s) > em^{\max}(s) \tag{38}$$

After updating the penalty factor using Equation (38), optimization is performed again through the ABC algorithm.

Figure 3 summarizes the proposed CEED solution. The procedure is performed sequentially as follows:

Step 1: Construct the MG model comprising DG, PV, and WT units and controllable demand, as shown in Figure 1, and initialize the input data for Equations (1)–(9).

Step 2: Establish the CEED problem and constraints according to Equations (10)–(26).

Step 3: Curtail the peak demand using EBDR as a base for elasticity from Equations (27)–(33).

Step 4: Set up parameters of the ABC algorithm, including the number of populations, maximum cycle, and maximum exploration.

Step 5: Initialize the power generation capacity of each unit randomly based on the initialization step in Equation (35).

Step 6: Search the neighborhood of the refence position through the employed bees, using Equation (36).

Step 7: Subject to Equation (37), apply a greedy search and choose one position to subsequently search for new neighboring positions.

Step 8: Replace the abandoned solution with the new solution found by the scout bees.

Step 9: Check that the result does not violate the emission constraint.

Step 10: If the emission constraint is violated, adjust the penalty factor in Equation (38).

Step 11: Repeat this process until the MG has satisfied all constraints.

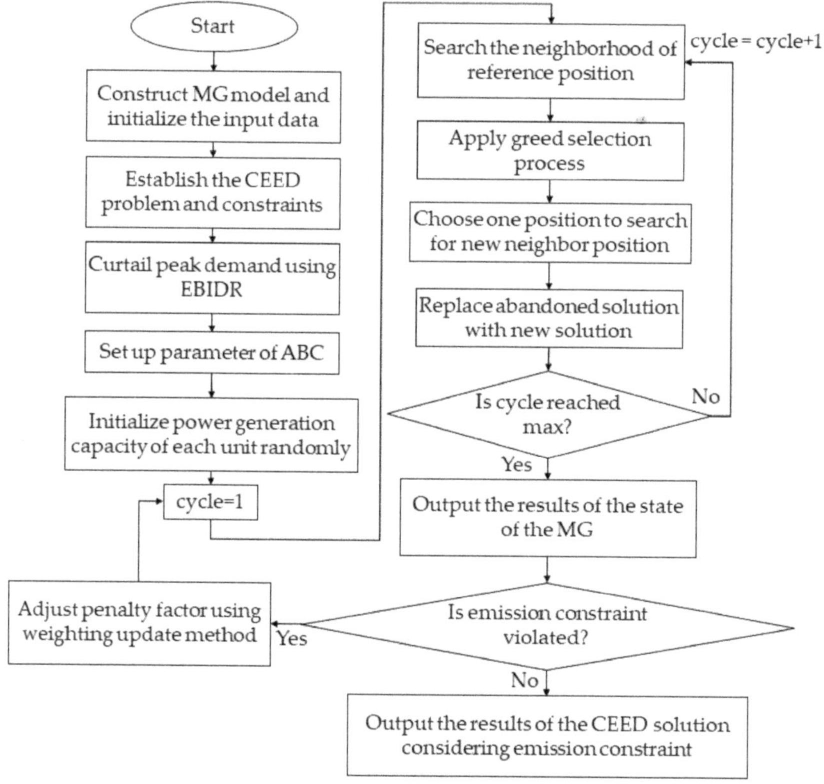

Figure 3. Overall process of the combined economic emission dispatch (CEED) solution.

6. Simulation Results

6.1. System Data

To verify the effectiveness and validity of the proposed approach, a simulation is conducted on the modified grid-connected MG system consisting of three DG units, one 40-MW PV system, and one 30-MW wind farm [25]. Table 1 lists each DG characteristic, including the min/max power limit, generation costs, and emission coefficients [32]. The operation costs for the wind farm and PV system include maintenance and investment costs of 30.8\$/MWh and 23.4\$/MWh, respectively [32].

Table 1. Diesel generation (DG) profiles.

DG Source	Min Power (MW)	Max Power (MW)	a_i ($\$/MW^2h$)	b_i (\$/MWh)	c_i (\$/h)	α_i (kg/MW^2h)	β_i (kg/MWh)	γ_i (kg/h)
G1	30	120	0.024	21	0	0.0105	−1.355	60
G2	32	128	0.029	20.16	0	0.008	−0.6	45
G3	40	152	0.021	20.4	0	0.012	−0.555	90

Table 2 depicts the hourly demand and generation power of RES, which are calculated for wind speeds and solar irradiation in the east coast of the USA [32]. Table 3 shows the market price and DR incentive over the 24-h period [38]. The pollutant emission violation fee is set to 6.34\$/kg [39]. The maximum DR capacity is assumed to be 10% of the total demand for each hour, while the maximum power flow capacity between MG and the main grid is 30 MW. The proposed WU-ABC algorithm has been applied to solve the associated CEED problem. To that end, we used MATLAB R2020a installed on a personal laptop with a 2.90 GHz core i5-9400F processor and 16 GB RAM. A simulation is run with 30 food sources and 300 iterations for 100 repeated trials, and the weighting update rate is set to 1.

Table 2. Demand and forecasted hourly output of RES.

Time (h)	Demand (MW)	PV (MW)	WT (MW)
1	140	0	1.7
2	150	0	8.5
3	155	0	9.27
4	160	0	16.66
5	165	0	7.22
6	170	0.03	4.91
7	175	6.27	14.66
8	180	16.18	25.56
9	210	24.05	20.58
10	230	39.37	17.85
11	240	7.41	12.80
12	250	3.65	18.65
13	240	31.94	14.35
14	220	26.81	10.35
15	200	10.08	8.26
16	180	5.30	13.71
17	170	9.57	3.44
18	185	2.31	1.87
19	200	0	0.75
20	240	0	0.17
21	225	0	0.15
22	190	0	0.31
23	160	0	1.07
24	145	0	0.58

Table 3. Market price and DR incentive.

Time (h)	Market Price (MW)	DR Incentive (MW)	Time (h)	Market Price (MW)	DR Incentive (MW)
1	30.70	27.30	13	69.90	35.50
2	25.70	26.70	14	64.10	38.80
3	21.40	26.50	15	60.00	38.10
4	17.30	28.20	16	52.00	38.10
5	14.90	27.50	17	39.10	37.50
6	16.60	27.60	18	32.20	36.50
7	16.10	29.70	19	29.50	35.70
8	16.60	29.20	20	27.20	34.50
9	26.40	30.10	21	15.20	33.80
10	32.70	31.90	22	12.10	29.80
11	36.50	33.80	23	12.80	28.70
12	56.90	33.90	24	20.20	29.60

6.2. Comparison Analysis

The U.S. plans to reduce CO_2 emissions from fossil fuel power plants by 32% before 2030 [39]. Thus, in this study, we assumed that emissions should ideally reduce by 32% when compared with an optimal operation situation of DG without consideration of emissions. Table 4 shows the optimal operation results when operating the MG with only DG units. The total operation costs and pollutant emissions are \$101,065.60 and 5064.76 kg, respectively. Thus, maximum emission is set at 3444.04 kg, which is 32% of the pollutant emissions from DG units.

Table 4. Results for DG.

	G1	G2	G3	Utility	Total
Operation costs (\$)	30,381.55	32,299.94	41,825.78	−3441.64	101,065.60
Pollutant emissions (kg)	789.95	1159.43	3115.38	0	5064.76

To reduce operation costs and to avoid the violation of the maximum emission constraint, the proposed EBDR and WU-ABC algorithms are used for solving the CEED problem. The penalty factor should be set taking generation costs and the emissions of each generator into account. Various types of penalty factors are calculated and listed in Table 5. These factors vary according to the characteristics of each generator. In multiobjective problems, the penalty factor affects the relative importance of different objective functions, because it affects the optimal solution. In other words, a penalty factor that is excessively large may overestimate the effect of one objective function in a multiobjective problem. Thus, a min-max relationship with a small average penalty factor is chosen to prevent predominant effects of only one objective function, and the penalty factor is iteratively updated from the initial value to find the best solution that satisfies all constraints.

Table 5. Various price penalty factors of each DG unit.

Penalty Factor Type	Formulation	h_1 (\$/kg)	h_2 (\$/kg)	h_3 (\$/kg)	h_{avg} (\$/kg)
Max-Min	$\dfrac{a_i\left(P_{G,i}^{max}\right)^2+b_i P_{G,i}^{max}+c_i}{\alpha_i\left(P_{G,i}^{min}\right)^2+\beta_i P_{G,i}^{min}+\gamma_i}$	99.50	89.89	41.22	76.87
Max-Max	$\dfrac{a_i\left(P_{G,i}^{max}\right)^2+b_i P_{G,i}^{max}+c_i}{\alpha_i\left(P_{G,i}^{max}\right)^2+\beta_i P_{G,i}^{max}+\gamma_i}$	58.96	30.78	12.68	34.14
Min-Min	$\dfrac{a_i\left(P_{G,i}^{min}\right)^2+b_i P_{G,i}^{min}+c_i}{\alpha_i\left(P_{G,i}^{min}\right)^2+\beta_i P_{G,i}^{min}+\gamma_i}$	22.63	19.85	9.77	17.41
Min-Max	$\dfrac{a_i\left(P_{G,i}^{min}\right)^2+b_i P_{G,i}^{min}+c_i}{\alpha_i\left(P_{G,i}^{max}\right)^2+\beta_i P_{G,i}^{max}+\gamma_i}$	13.41	6.80	3.00	7.74

In this study, three cases are considered to validate the proposed approach. Case 1 is a conventional CEED proposition without a DR program and penalty factor update. Case 2 involves a conventional economic DR program but does not include penalty factor updates. In Case 3, the proposed EBDR is used and the penalty factor is updated through the WU-ABC algorithm. All cases are simulated under the same constraints, except for the DR programs and penalty factor update.

Figure 4 shows the hourly scheduling of optimal power generation for all cases. To reduce operation costs, the MGO sells power for utility when market prices are high, in all cases. Although the generation cost of G1 is higher than that of the two other generators, it generates the most power due to low pollutant emissions per MWh. As shown in Figure 4a, Case 1 does not shift the peak demand; operation mainly occurs through the power generated by DG units, because no DR program is considered. Compared to Case 1, Case 2 utilizes a conventional economic DR program and demonstrates demand shifts in consideration of elasticity according to market prices. As shown in Figure 4b, the MGO uses the DR program to shift demand from periods of high market price to periods when power is relatively inexpensive. Thus, the power generation of DG units in the MG is reduced, and the operation costs are reduced despite considering incentives for the DR participants. However, in Case 2, no maximum emission constraints are imposed; to remedy this problem, penalty factor update and EBDR are considered in Case 3. The results for Case 3 are shown in Figure 4c. Compared to Case 1 and Case 2, it can be noted that the period taken by the MGO to sell power is reduced and the number of participants in the DR program is increased. Moreover, the power generation capacity of G2 and G3 (which have relatively large pollutant emissions per unit power generated) is also decreased. Case 3 could satisfy the maximum emission constraint.

Table 6 shows the optimal CEED results for each case. Compared to the results in Table 4, which are essentially optimal operation propositions without environmental considerations, it can be seen that Case 1 has marginally higher operation costs (by 1.1%) but shows significant reduction in pollutant emissions (by 26.1%). These results violate the maximum emission constraint and there is no significant reduction in operation costs. For Case 2, the MGO reduced the operation costs of DG by using an economic DR to reduce peak demand. Thus, additional DR incentive costs are incurred, but the total operation costs are reduced by selling more power during high market price periods. The operation costs are reduced to \$98,187.99, and pollutant emissions are decreased to 3571.14 kg. These results show that scheduling under a DR program for MG operation can have a positive impact, leading to reductions in peak demand, operating costs, and pollutant emissions. Compared to Case 1, Case 2 reduces both operation costs and pollutant emissions, but it does not satisfy the maximum emission constraints. Therefore, the MGO pays \$805.81 as a violation fee, which is proportional to the excess pollutant emissions. In Case 3, the proposed EBDR and penalty factor update method are used to satisfy the maximum emissions constraint. Compared to Case 2, the operation costs of each DG unit are reduced due to the reduction of the power generation capacity of the units. The quantity of power that is sold to the utility decreased. The cost of participating in the DR program increased with the use of EBDR. Consequently, operating costs increased by approximately 0.05% compared to Case 2, but the maximum emission constraint is satisfied, curtailing emissions within 3444.03 kg. As can be seen from these results, the proposed approach is effective in solving the CEED problem with a maximum emission constraint.

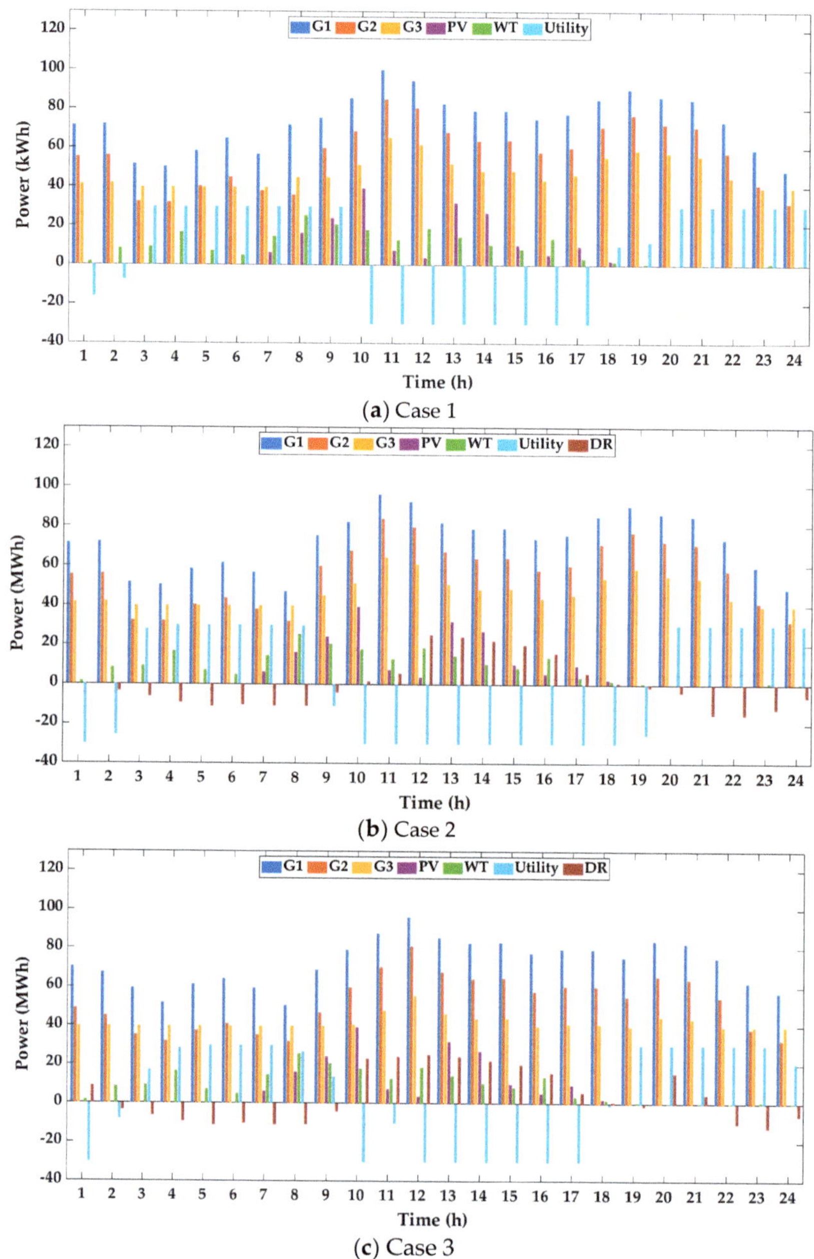

Figure 4. Optimal hourly scheduling in the MG for each case.

Table 6. Operation costs and pollutant emissions in each case.

Case		G1	G2	G3	PV	WT	Utility	DR	Total
1	Operation costs ($)	40,247.6	30,878.3	25,262.5	5635.5	4992.9	−3850.2	-	102,166.6
	Emissions (kg)	484.63	947.20	2207.49	0	0	0	-	3739.32
2	Operation costs ($)	39,529.5	29,734.6	24,149.9	5635.5	4992.9	−10,246.8	4392.5	98,188.1
	Emissions (kg)	457.29	928.02	2185.83	0	0	0	0	3571.14
3	Operation costs ($)	39,475.1	27,264.8	21,502.9	5635.5	4992.9	−6049.2	5417.8	98,239.8
	Emissions (kg)	442.52	888.15	2113.36	0	0	0	0	3444.03

Figure 5 indicates the penalty factor and pollutant emissions with respect to the number of updates. The penalty factor is updated proportionally to the excess pollutant emissions; consequently, the penalty factor update is performed 11 times, and the penalty factor is finally determined to be 23.55$/kg. Table 7 summarizes the optimal CEED solution for all cases. Case 1 without a DR program presents a greater total operation cost and higher emissions than the other cases. Meanwhile, for Case 2 with the conventional economic DR program, the MGO is able to reduce operation costs and pollutant emissions relative to Case 1 but pays a violation fee for exceeding the maximum emission constraint. In Case 3, the maximum emission constraint is satisfied through the use of penalty factor update and the EBDR program; although it slightly increases the operating cost compared to Case 2, the constraint violation cost of $805.81 is not incurred. Therefore, it can be concluded that the proposed approach exhibits the most promising results for the CEED problem, when considering a violation fee.

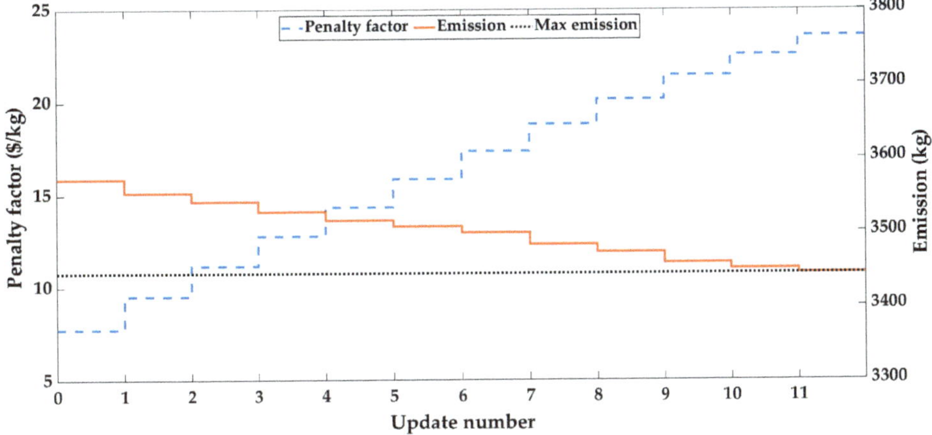

Figure 5. Penalty factor update and emissions.

Table 7. Summary of results for all cases.

	Case 1	Case 2	Case 3
Total operation costs ($)	105,766.50	98,187.99	98,239.85
Total emissions (kg)	3739.32	3571.14	3444.03
Violation fee ($)	1872.08	805.81	0

To demonstrate the superiority of WU-ABC, performance tests have been conducted through comparison with various algorithms [40,41]. For a fair comparison, the maximum emission constraints and violation fee are equally considered in grid-connected MG operation. A total of five algorithms are considered, such as ABC, improved multilayer artificial bee colony (IML-ABC) [41], weighting update genetic algorithm (WU-GA), weighting update particle swarm optimization (WU-PSO), and proposed WU-ABC. WU-GA and WU-PSO are the addition of the weighting update method to the conventional

GA and PSO. The simulation process for all algorithms is repeated 10 times, and then the results are calculated as average value. Table 8 shows the CEED results of each algorithm. It can be observed that WU-ABC reaches a best solution compared with other algorithms. The algorithms without the weighting update method do not satisfy the emission constraint, and as a result, the total cost is high due to imposing violation fees. In addition, WU-ABC indicates the lowest cost compared to other algorithms considered weighting update methods and takes the least CPU time. Therefore, these results reveal that the proposed WU-ABC algorithm is appropriate for solving CEED problems considering maximum emission constraints.

Table 8. Comparison results for various algorithms.

Algorithm	Cost ($)	Emission (kg)	CPU Time (sec)
ABC	98,993.80	3571.14	29.8
IML-ABC	98,792.46	3568.27	24.7
WU-GA	98,411.38	3444.04	26.5
WU-PSO	98,408.75	3444.04	25.9
WU-ABC	98,239.85	3444.03	18.3

7. Conclusions

This paper proposed a CEED approach with the EBDR program by using the WU-ABC algorithm. The objective function was constructed in consideration of generation costs and pollutant emission. The emission function took the emitted quantities of CO_2, SO_x, and NO_x, into account and converts their sum into operation costs through unit conversion, using a penalty factor. To meet the maximum emission constraint considered in addition to the CEED in our work, the EBDR program was proposed to balance economics and environmental issues. The EBDR program is based on the concept of elasticity, and the participation volume changes depending on the violation of emission constraints. To solve the multiobjective optimization problem, we proposed the use of the WU-ABC algorithm, a combination of the weighting update method and the conventional ABC algorithm. To demonstrate the effectiveness of the proposed CEED approach, simulations were conducted on the modified grid-connected MG system. Three cases were considered depending on the maximum emission constraint and the application of a DR program. The proposed approach was able to reduce operation costs compared to the conventional solution without DR program. Furthermore, it was easy to satisfy the maximum emission constraints by updating the penalty factor. In summary, the use of the EBDR program reduced operation costs and pollutant emissions, and the maximum emission constraints were satisfied through the WU-ABC algorithm. Although operation costs were slightly increased compared to that of a conventional economic DR, an emission violation fee was not imposed. In addition, the performance test results indicated that the WU-ABC algorithm outperformed other algorithms in terms of cost and CPU time. Therefore, the proposed CEED approach can assist MGOs in optimizing MG operations under environmental constraints. Our future work will be under way to focus on solving the sensitivity problem in MG when the consumers do not comply with the command of MGO in the CEED problem.

Author Contributions: H.-S.R. proposed the main idea of this paper and M.-K.K. coordinated the proposed approach and thoroughly reviewed the manuscript. All authors read and approved the manuscript.

Funding: This research was supported by the Korea Electric Power Corporation (grant numbers: R18XA06-75). This research was also supported by Basic Science Research Program through the National Research Foundation of Koera (NRF) funded by the Ministry of Education (2020R1A2C1004743).

Conflicts of Interest: The authors declare no conflict of interest.

Nomenclature

Constants

a_i, b_i, c_i, d_i	Fuel cost coefficients of i DG
SI	solar irradiation
a, b	Shape parameters of beta PDF
μ_s	Mean of solar irradiation data
σ_s	Standard deviation of solar irradiation data
η_j	Efficiency of j PV panel
A_j	Area size of j PV panel
β	Temperature coefficients of maximum power of PV panel
d, C	Shape parameters of Weibull PDF
P_r	Rated power of WT unit
v_{ci}	Cut-in wind speed of WT unit
v_{co}	Cut-out wind speed of WT unit
v_r	Rated wind speed of WT unit
I	Number of DG units
J	Number of PV panels
K	Number of WT units
t	Time
$AC_{s,j}$	Annuitization coefficient of j PV panels
$AC_{w,k}$	Annuitization coefficient of k WT units
$I_{s,j}$	Maintenance cost of j PV panels
$I_{w,k}$	Maintenance cost of k WT units
$G_{s,j}$	Installation cost of j PV panels
$G_{w,k}$	Installation cost of k WT units
$\alpha_i, \beta_i, \gamma_i, \zeta_i, \lambda_i$	i DG emission coefficient
L	Number of transmission lines
h	Penalty factor

Variables

$C_{G,i}(t)$	Generation cost function of i DG units at t
$P_{G,i}(t)$	Generation power of i DG units at t
$P_{G,i}^{\min}(t)$	Minimum generation power of i DG units at t
$P_{G,i}^{\max}(t)$	Maximum generation power of i DG units at t
PDF_B	Beta PDF
Γ	Gamma function
T_t	Temperature at t
$P_{s,j}(t)$	Generation power of j PV panels at t
$P_{s,j}^{\min}(t)$	Minimum generation power of j PV panels at t
$P_{s,j}^{\max}(t)$	Maximum generation power of j PV panels at t
PDF_w	Weibull PDF
v_t	Wind speed at t
$P_{w,k}(t)$	Generation power of k WT units at t
$P_{w,k}^{\min}(t)$	Minimum generation power of k WT units at t
$P_{w,k}^{\max}(t)$	Maximum generation power of k WT units at t
$P_u(t)$	Power exchanged with utility at t
$P_u^{\min}(t)$	Minimum power exchanged with utility at t
$P_u^{\max}(t)$	Maximum power exchanged with utility at t
$P_{buy}(t)$	Power purchased from utility at t
$P_{sell}(t)$	Power sold to utilities at t
$b_u(t)$	Unit function
$C_u(t)$	Power exchanged cost with utility at t

$\lambda_{sell}(t)$	Power purchased cost from utility at t
$\lambda_{buy}(t)$	Power sold costs to utilities at t
$F(P)$	Generation costs function
$C_{s,j}$	Solar power cost
$C_{w,k}$	Wind power cost
$E(P)$	Emission function
$em(t)$	Pollutant emissions at t
$em^{\max}(t)$	Maximum pollutant emissions at t
$C(P)$	CEED function
$C_{DR}(t)$	Total DR cost at t
$I(\Delta D(t))$	Total Incentive cost
$P(\Delta D(t))$	Total Penalty cost
$P_D(t)$	Total power demand at t
$P_{loss}(t)$	Total line loss at t
$D_0(t)$	Initial power demand at t
$D(t)$	Power demand at t
$\Delta D(t)$	Change in power demand
S_l	Power flow at line l
$S_l^{\max}$	Maximum power flow at line l
$E(t_1, t_2)$	Elasticity for t_1 and t_2
$MP_0(t)$	Initial market price at t
$MP(t)$	Market price at t
$inc(t)$	DR incentive cost at t
$pen(t)$	DR penalty cost at t
$CP(t)$	Contracted power capacity at t
$D_e(t)$	Power demand considering emission
$E_e(t_1, t_1)$	Emission elasticity
x_{mn}	m^{th} random food source in dimension n
$x_{mn}^{\min}$	Minimum m^{th} random food source in dimension n
$x_{mn}^{\max}$	Maximum m^{th} random food source in dimension n
v_{mn}	Best position
ϕ_{mn}	Random number in the range $[-1, 1]$
P_m	Probability of m food source being chosen
fit_m	Fitness function

References

1. Zhou, X.; Ai, Q.; Yousif, M. Two kinds of decentralized robust economic dispatch framework combined distribution network and multi-microgrids. *Appl. Energy* **2019**, *253*, 113588. [CrossRef]
2. Huijun, L.; Yungang, L.; Fengzhong, L.; Yanjun, S. A multiobjective hybrid bat algorithm for combined economic/emission dispatch. *Int. J. Electr. Power Energy Syst.* **2018**, *101*, 103–115.
3. Ghorab, M. Energy hubs optimization for smart energy network system to minimize economic and environmental impact at Canadian community. *Appl. Therm. Eng.* **2019**, *151*, 214–230. [CrossRef]
4. Ioanna, P.; Voula, P.K. The impact of carbon emission fees on passenger demand and air fares: Game theoretic approach. *J. Air Transp. Manag.* **2016**, *55*, 41–51.
5. Hossein, N.; Seyed-Ehsan, R.; Ali, A.; Ehsan, N.; Mehdi, F.; Mohammad, H.A.; Mohammad, R.N. A multi-objective framework for multi-area economic emission dispatch. *Energy* **2018**, *154*, 126–142.
6. Dhillon, J.S.; Parti, S.C.; Kothari, D.P. Stochastic economic emission load dispatch. *Electr. Power Syst. Res.* **1993**, *26*, 179–186. [CrossRef]
7. Kularni, P.S.; Kothari, A.G.; Kothari, D.P. Combined economic and emission dispatch using improved backpropagation neural network. *Electr. Mach.* **2000**, *28*, 31–44.
8. Majid, M.; Sayyad, N.; Kazem, Z. A cost-emission framework for hub energy system under demand response program. *Energy* **2017**, *134*, 157–166. [CrossRef]

9.   Tong, X.; Hongyu, L.; Zhongfu, T.; Liwei, J. Coordinated energy management for micro energy systems considering carbon emissions using multi-objective optimization. *Energies* **2019**, *12*, 4414.

10.   Albadi, M.H.; El-saadany, E.F. A summary of demand response in electricity market. *Electr. Power Syst. Res.* **2008**, *78*, 1989–1996. [CrossRef]

11.   Kim, H.J.; Kim, M.K. Multi-Objective Based optimal energy management of grid-connected microgrid considering advanced demand response. *Energies* **2019**, *12*, 4142. [CrossRef]

12.   Che, L.; Zhang, X.; Shahidehpour, M.; Alabdulwahab, A.; Abusorrah, A. Optimal interconnection planning of community microgrids with renewable energy sources. *IEEE Trans. Smart Grid* **2017**, *8*, 1054–1063. [CrossRef]

13.   Nwulu, N.I.; Xia, X. Optimal dispatch for microgrid incorporating renewables and demand response. *Renew. Energy* **2017**, *101*, 16–28. [CrossRef]

14.   El-keib, A.A.; Ma, H.; Hart, J.L. Environmentally constrained economic dispatch using a lagrangian relaxation method. *IEEE Trans. Power Syst.* **1994**, *9*, 1723–1729. [CrossRef]

15.   Basu, A.K.; Bhattacharya, A.; Chowdhury, S. Planned scheduling for economic power sharing in a CHP-based micro-grid. *IEEE Trans. Power Syst.* **2012**, *27*, 30–38. [CrossRef]

16.   Moghaddam, A.A.; Seifi, A.; Niknam, T.; Alizadeh Pahlavani, M.R. Multi-objective operation management of a renewable MG (micro-grid) with back-up micro turbine/fuel cell/battery hybrid power source. *Energy* **2011**, *36*, 6490–6507. [CrossRef]

17.   Ghasemi, M.; Aghaei, J.; Akbari, E.; Ghavidel, S.; Li, L. A differential evolution particle swarm optimizer for various types of multi-area economic dispatch problems. *Energy* **2016**, *107*, 182–195. [CrossRef]

18.   Jacob Rahlend, I.; Veeravalli, S.; Sailaja, K.; Sudheera, B.; Kothari, D.P. Comparison of AI techniques to solve combined economic emission dispatch problem with line flow constraints. *Int J. Electr. Power Energy Syst.* **2010**, *32*, 592–598. [CrossRef]

19.   Aghajani, G.R.; Shayanfar, H.A.; Sshayeghi, H. Presenting a multi-objective generation scheduling model for pricing demand response rate in micro-grid energy management. *Energy Convers. Manag.* **2015**, *106*, 308–321. [CrossRef]

20.   Nojavan, S.; Majidi, M.; Zare, K. Optimal scheduling of heating and power hubs under economic and environment issues in the presence of peak load management. *Energy* **2018**, *156*, 34–44. [CrossRef]

21.   Nwulu, N.; Xia, X. Multi-objective dynamic economic emission dispatch of electric power generation integrated with game theory based demand response programs. *Energy Convers. Manag.* **2015**, *89*, 963–974. [CrossRef]

22.   Behrangrad, M.; Sugihara, H.; Funaki, T. Effect of optimal spinning reserve requirement on system pollution emission considering reserve supplying demand response in the electricity market. *Appl. Energy* **2011**, *88*, 2548–2558. [CrossRef]

23.   Parvania, M.; Fotuhi-Firuzabad, M. Demand response scheduling by stochastic SCUC. *IEEE Trans. Smart Grid* **2010**, *1*, 89–98. [CrossRef]

24.   Abdelaziz, A.Y.; Ali, E.S.; Abd Elazim, S.M. Combined economic and emission dispatch solution using flower pollination algorithm. *Int. J. Electr. Power Energy Syst.* **2016**, *80*, 264–274. [CrossRef]

25.   Dey, B.; Roy, S.K.; Bhattacharyya, B. Solving multi-objective economic emission dispatch of a renewable integrated microgrid using latest bio-inspired algorithms. *Eng. Sci. Technol. Int. J.* **2019**, *22*, 55–66. [CrossRef]

26.   Guvenc, U.; Sonmez, S.; Duman, S.; Yorukeren, N. Combined economic and emission dispatch solution using gravitational search algorithm. *Sci. Iran.* **2012**, *19*, 1754–1762. [CrossRef]

27.   Koltsaklis, N.E.; Giannakakis, M.; Georgiadis, M.C. Optimal energy planning and scheduling of microgrids. *Chem. Eng. Res. Des.* **2018**, *131*, 318–332. [CrossRef]

28.   Mengelkamp, E.; Gärttner, J.; Rock, K.; Kessler, S.; Orsini, L.; Weinhardt, C. Designing microgrid energy markets: A case study: The Brooklyn Microgrid. *Appl. Energy* **2018**, *210*, 870–880. [CrossRef]

29.   Lu, X.; Liu, Z.; Ma, L.; Wang, L.; Zhou, K.; Feng, N. A robust optimization approach for optimal load dispatch of community energy hub. *Appl. Energy* **2020**, *259*, 114195. [CrossRef]

30.   Chiang, C.L. Improved genetic algorithm for power economic dispatch of units with valve-point effects and multiple fuels. *IEEE Trans. Power Syst.* **2005**, *20*, 1690–1699. [CrossRef]

31.   Sampaio, P.G.V.; Gonzalez, M.O.A. Photovoltaic solar energy: Conceptual framework. *Renew. Sustain. Energy Rev.* **2017**, *74*, 590–601. [CrossRef]

32. Augustine, N.; Suresh, S.; Moghe, P.; Sheikh, K. Economic dispatch for a microgrid considering renewable energy cost functions. In Proceedings of the IEEE PES Innovative Smart Grid Technologies (ISGT), Washington, DC, USA, 16–20 January 2012; pp. 1–7.
33. Devi, A.L.; Krishna, O.V. Combined economic and emission dispatch using evolutionary algorithms-a case study. *ARPN J. Eng. Appl. Sci.* **2008**, *3*, 28–35.
34. Moghaddam, M.P.; Abdollahi, A.; Rashidinehad, M. Flexible demand response programs modeling in competitive electricity markets. *Appl. Energy* **2011**, *88*, 3257–3269. [CrossRef]
35. Shahryari, E.; Shayeghi, H.; Mohammadi-Ivatloo, B.; Moradzadeh, M. An improved incentive-based demand response program in day-ahead and intra-day electricity markets. *Energy* **2018**, *155*, 205–214. [CrossRef]
36. Secui, D.C. A new modified artificial bee colony algorithm for the economic dispatch problem. *Energy Convers. Manag.* **2015**, *89*, 43–62. [CrossRef]
37. Paliwal, N.K.; Singh, A.K.; Singh, N.K. Energy scheduling optimisation of an islanded microgrid via artificial bee colony guided by global best, personal best and asynchronous scaling factors. *Int. J. Sustain. Energy* **2020**, *39*, 539–555. [CrossRef]
38. Rezaee, J.A. Dynamic environmental-economic load dispatch in grid-connected microgrids with demand response programs considering the uncertainties of demand, renewable generation and market price. *Int. J. Numer. Model. Electron. Netw. Devices Fields* **2020**, e2798. [CrossRef]
39. Hafstead, M. Projected CO_2 Emissions Reductions under the American Opportunity Carbon Fee Act of 2017. *Resources for the Future Issue Brief.* 2017, 17-09. Available online: https://www.rff.org/ (accessed on 1 November 2020).
40. Hussain, S.; Al-Hitmi, M.; Khaliq, S.; Hussain, A.; Asghar Saqib, M. Implementation and comparison of particle swarm optimization and genetic algorithm techniques in combined economic emission dispatch of an independent power Plant. *Energies* **2019**, *12*, 2037. [CrossRef]
41. Ryu, H.S.; Kim, M.K. Two-Stage Optimal Microgrid Operation with a Risk-Based Hybrid Demand Response Program Considering Uncertainty. *Energies* **2020**, *13*, 6052. [CrossRef]

Publisher's Note: MDPI stays neutral with regard to jurisdictional claims in published maps and institutional affiliations.

Article

Relieving Tensions on Battery Energy Sources Utilization among TSO, DSO, and Service Providers with Multi-Objective Optimization

Gianni Celli, Fabrizio Pilo, Giuditta Pisano *, Simona Ruggeri and Gian Giuseppe Soma

Department of Electric and Electronical Engineering, University of Cagliari, 09123 Cagliari, Italy; gcelli@unica.it (G.C.); farizio.pilo@unica.it (F.P.); simona.ruggeri@unica.it (S.R.); giangiuseppe.soma@unica.it (G.G.S.)
* Correspondence: giuditta.pisano@unica.it; Tel.: +39-070-675-5868

Received: 31 October 2020; Accepted: 31 December 2020; Published: 5 January 2021

Abstract: The European strategic long-term vision underlined the importance of a smarter and flexible system for achieving net-zero greenhouse gas emissions by 2050. Distributed energy resources (DERs) could provide the required flexibility products. Distribution system operators (DSOs) cooperating with TSO (transmission system operators) are committed to procuring these flexibility products through market-based procedures. Among all DERs, battery energy storage systems (BESS) are a promising technology since they can be potentially exploited for a broad range of purposes. However, since their cost is still high, their size and location should be optimized with a view of maximizing the revenues for their owners. Intending to provide an instrument for the assessment of flexibility products to be shared between DSO and TSO to ensure a safe and secure operation of the system, the paper proposes a planning methodology based on the non-dominated sorting genetic algorithm-II (NSGA-II). Contrasting objectives, as the maximization of the BESS owners' revenue and the minimization of the DSO risk inherent in the use of the innovative solutions, can be considered by identifying trade-off solutions. The proposed model is validated by applying the methodology to a real Italian medium voltage (MV) distribution network.

Keywords: distribution network planning; energy storage system; multi-objective optimization; optimal location; risk assessment; flexibility; distributed energy resources; distribution system operators; local services; system services; arbitrage; frequency control

---

1. Introduction

The production of electric energy with renewable energy sources (RES) and the electrification of energy use are crucial for energy transition, which is the ongoing process to reduce the use of fossil fuels with the decarbonization, decentralization, and digitalization of the energy sectors. In this context, the European Union (EU) launched an ambitious plan to cut emissions in the atmosphere by harvesting the energy from wind and solar and fostering a profound transformation of heating and transportation sectors [1].

From the power system perspective, the non-programmability of energy production (i.e., mostly based on intermittent sources) is making it tough to comply with the required adequacy and security levels. Furthermore, since a significant amount of renewable energy currently comes from small to medium-size power plants connected to the distribution system (e.g., photovoltaic power plants), the energy transition is profoundly impacting the distribution system. The high share of distributed generation (DG) is already causing temporary voltage regulation and power congestion problems in distribution networks that are destined to become more frequent with the progress of the energy transition. The active management (AM) of distribution networks allows fixing such operational issues with the use of flexibility offered by generation and consumption, that is referred to as no-network solutions in opposition to the traditional paradigm of network solutions. As a consequence, no-network options have to be included in distribution planning as a new development option [2–4]. The usage of flexibility can increase the hosting capacity of the network and, as a consequence, the deferment of infrastructural investments. To foster the use of flexibility, the EU Member States now have to allow the distribution system operators (DSOs) to procure flexibility products with transparent, non-discriminatory, and market-based procedures for the operation and development of the distribution system [5]. Because of the expected more extensive exploitation of flexibility for the operation of the distribution network, DSO and transmission system operators (TSO) will have to cooperate, by exchanging all necessary information to ensure the use of flexibility, guarantee an efficient operation of the whole system, and facilitate market development [6]. Indeed, the "local" flexibility, offered by the distributed energy resources (DERs) for the operation of the distribution system, might be useful also for "system" services, needed by the TSO for guaranteeing safe and secure operation of the whole power system. It is clear that DSOs cannot completely rely on the flexibility in their networks, and the level of uncertainty might be considerably high.

Battery energy storage systems (BESS) are a technical choice to increase flexibility and reduce the level of uncertainty. They can be locally employed for RES integration, load peak flattening, voltage regulation, efficiency improvement, solving power congestions, etc., BESS can offer frequency regulation services that are crucial with the progressive drop of the system inertia consequent to the diffusion of non-rotating generators or power electronics interfaced generators. The reduction of the system inertia may cause critical frequency deviations even during small active power imbalances [7].

BESS can foster the market of flexibility products in the early stage of implementation. In fact, pioneering projects implementing flexibility markets involving both TSOs and DSOs showed that the sourcing of flexibility is a critical element, mostly at the early phases of the market development, when aspects like flexibility product cost, provider participation, and availability of flexibility products represent a potential risk for the distribution network management. This uncertainty could be reduced by BESS that represent an additional resource when the number of available RES is too low to ensure effective competition in the area where suitable facilities for the provision of the services are located, avoiding the increase of the costs.

Moreover, since also TSOs are procuring the flexibility in the market, in the paper, it is proposed a decision-making process that allows the evaluation of the share of flexibility to be reserved to each player, enabling a stronger coordination between local and system objectives and needs. The BESS cost is obviously a crucial element of the decision-making process that has been considered in the paper. In fact, despite BESS having a huge cost-reduction potential, BESS cost is still a feature that impacts on the success of business models depending on the regulatory framework. In [8], the authors proposed a multi-objective (MO) approach for simultaneously optimizing the size, the position, and the operation profile of BESS called to offer services to the DSOs, without monetizing the benefits that are not naturally expressed with currency (e.g., improvement of voltage profile or benefits related to the environmental protection). That paper considered the point of view of the DSOs that, under specific conditions, can be allowed to own and manage BESS. This paper makes another step forward, combining the needs of BESS owners and DSOs. Taking into account the advocated cooperation between TSO and DSO, an instrument is proposed that allows the evaluation of the availability of the flexibility product for both the systems' operators, without disregarding the BESS owner point of view. The BESS owners aim at increasing their incomes from arbitrage practice and from the provision of ancillary services to both the system operators (i.e., DSO and TSO), while the DSO aims at reducing the residual risk of technical constraints violation by promoting the use of flexibility products offered by the BESS owners. In particular, for considering both the interests, the capacity and the power of the BESS are shared between the percentage dedicated exclusively to support the DSO operation and the remaining quote available for arbitrage or other services. The specific service of frequency regulation support offered by BESS to the TSO and the relevant economic benefits for the BESS owners are explicitly considered in the proposed optimization.

The paper is organized as follows. In Section 2, the proposed methodology is described underlying the novelties and the improvements in respect to the one proposed in [8]. In Section 3, the approach is validated and discussed through a case study that applies the methodology to a real Italian distribution network. Finally, in Section 4, some concluding remarks are presented.

2. Multi-Objective Optimization for Optimal Exploitation of BESS

A large number of papers have been published on the use of BESS in power distribution systems, analyzing different models and methods to enhance the optimal network planning [9–11]. Studies that include BESS (as well as demand response actions, characterized by possible recover of the curtailed energy) are more complicated due to time intercorrelations, since the BESS energy scheduling in one hour is subject to the charging/discharging cycle implemented in the previous hours [12]. Consequently, the decision-making problem may become complicated to be solved because planning alternatives can excessively grow in number. Traditional numerical methods like non-linear programming, dynamic programming, and mixed-integer linear programming have shortcomings if applied on large and complex distribution systems. On the contrary, meta-heuristic algorithms, like particle swarm (PS), Tabu search (TS), evolutionary algorithm (EA), and genetic algorithm (GA), can provide near-optimal solutions for complex, large-scale planning problems, like the one faced in this paper [9]. In [13–15], the PS optimization is used for the optimal allocation of BESS in the distribution system. In [16], the use of an EA for determining the capacity of BESS in an islanded microgrid, considering both steady-state and dynamic constraints, is proposed. The problem is formulated as an MO optimization that involves the dynamic equations of the power system, to improve reliability, stabilizing transients and reducing load shedding. A significant number of papers used the GA [17] or a combination of GA with other optimization techniques like PS, linear programming in [18] or quadratic programming in [19].

It is worth noting that, despite the high interest related to the cooperation between DSO and TSO in the provision of the ancillary services, in literature, few publications have addressed the optimal size and location of BESS for the assessment of flexibility products to be shared between DSO and TSO. Moreover, most of them analyze the voltage support and peak shaving, while very few analyze the possible frequency service [20]. Regarding frequency support, such papers are related to high voltage (HV) networks, islanded networks (microgrid or island) or examine the aggregation of DERs (usually represented as virtual power plant, VPP) [21–23]. For instance, in [23] clusters of electric vehicles (EVs) are grouped together as a VPP to provide fast frequency reserve service to the transmission system through the DSO whilst considering network unbalance. Most of the papers on optimal location and size analyze a specific voltage level or the two systems independently without considering the possible services for the other level and a wide range of contingencies (or the worst-case scenario) are considered [24].

Compared to the literature, the proposed paper presents an advancement for several reasons. First of all, moving in the direction of the cooperation between DSO and TSO, different grid services (i.e., arbitrage, frequency containment reserve (FCR), and automatic frequency restoration reserve (aFRR)), that medium voltage (MV) BESS can sell through the services market are considered. Moreover, the planning strategies are not developed taking into account only critical days but typical daily profiles, indicative of the seasonal behavior of loads and generators during a year, are used. Such choice is in agreement with the recommendations of the main international scientific organizations (CIGRE, CIRED, IEEE, etc.), that recognize the unsuitability of the traditional deterministic distribution planning approaches, based on the aim of fulfilling the extremely rare operating conditions, which could lead to an unsustainable amount of network investments. In addition, the methodology proposed is able to deal with all the uncertainties related to RES and to consider distinctly the risk associated to any planning decision.

Generally, the optimal BESS exploitation requires to simultaneously take account of multiple goals. Thus, another critical aspect for the optimal siting and sizing of BESS in a distribution network is related to the definition of a unique financial objective function, because some benefits are not directly monetizable without adopting subjective assumptions that can produce biased results. In this context, MO programming permits a more transparent and impartial decision process and can be used for financial purposes by decision-making teams of companies or for socio-economic studies by regulators for defining fair rules [16, 17,19]. Evolutionary algorithms are well suited to solve optimization problems that are characterized by many contrasting objectives. Differently by other more conventional optimization methods, like the weighted linear combination or the ε-constraint, that need to perform several separate runs, the evolutionary algorithms can simultaneously deal with many candidates that form sets of solutions (named population) and find the optimal ones belonging to the Pareto set by performing the algorithm one time only. Furthermore, the Pareto front shape and its continuity have a small impact on the evolutionary algorithms results [25].

For these reasons, this paper implements a full MO optimization procedure based on a real coded non-dominated sorting genetic algorithm-II (NSGA-II) algorithm. This methodology has been chosen among meta-heuristic algorithms because it is recognized to be an efficient and robust technique, capable of generating good trade-off solutions for a wide range of optimization problems [26]. The real codification has been implemented by the authors in [8] to better deal with the continuous nature of some variables in optimization problems that include BESS exploitation. The main novelties of this paper, compared to [8], are the definition of new objective functions (especially the risk of technical constraint violations in a given electric distribution network) and the relative adaptations of the solution coding. Indeed, in [8], only the DSO point of view was considered, and the MO approach was used to avoid the a priori monetization of the benefits and perform the cost-benefit analyses of BESS allocation plans proposed by the DSO. Instead, in this paper, two stakeholders have been considered, the DSO and the BESS private owners, in order to find trade-off solutions between their contrasting goals.

2.1. Fitness Function Assessment in the NSGA-II

A typical genetic algorithm simulates the mechanism of the natural evolution, by applying systematically the three genetic operators of *selection* (that gives more chance of reproduction to the better individuals), *crossover* (that generates offspring solutions by mixing the genetic characteristics of parents), and *mutation* (that implements random alterations of the genetic characteristics) to evolve the population towards the global optimum. Often, *elitism* (i.e., only the better solutions are considered in the evolution process) is implemented in the formation of the new population in order to increase the effectiveness of the optimization procedure and speed up the convergence of the algorithm.

Taking account of these aspects, the mechanism implemented for comparing two solutions and identifying the better one is crucial for any evolutionary algorithm. The classification procedure used by the NSGA-II algorithm is based on the definition of two attributes associated with each individual: the *non-domination rank* and the *crowding distance*. The *non-domination rank* organizes the candidates into subsets of individuals non-dominated by any other in the same subset (fronts of non-dominance). Therefore, the first non-dominated front is formed by all the individuals of the Pareto set for which no other solutions are either equal to or better than them on all of the search objectives, and it is marked with the lowest (best) rank. The other fronts are sorted depending on the number of subsets from which they are dominated: e.g., the second front is dominated only by the first and dominates all the remaining fronts, the third front is dominated only by the first two and so on. In order to guarantee the diversity in each front, a second attribute is introduced, the *crowding distance* (CD), based on the cardinality of the solution sets and their distance to the solution boundaries. It is calculated by summing the absolute normalized differences in the function values of two adjacent solutions; the sum is extended to each objective function (*OF*), as indicated in (1):

$$CD_i = \sum_{k=1}^{N_{OF}} \frac{OF_k^{i+1} - OF_k^{i-1}}{OF_k^{max} - OF_k^{min}},\tag{1}$$

where N_{OF} is the number of *OF*s and OF_k^i is the kth *OF* value of the ith generic individual. The higher the *CD* of a solution, the less crowded the corresponding area of the front and, hence, the finer its fitness value.

The application of these two attributes allows finding a better solution in a pairwise comparison, assuring the correct implementation of selection operator and elitism in the evolution of the population.

2.2. Solution Coding for the BESS Location Problem

The traditional binary coding makes the GAs particularly suitable for solving sizing and siting problems of different resources, like (for power distribution systems) capacitors, distributed generators, or BESS. However, in the case of storage devices, the effectiveness of their support to network operation depends also on their hourly state of charge (SoC). Therefore, the solution coding has to represent the BESS daily scheduling as well, so that the NSGA-II can optimize the design and normal operation of storage devices simultaneously, finding full compromise solutions among different point of views. Furthermore, as it is explained in the next paragraph, part of the BESS capacity is reserved to DSO for local distribution network support. Consequently, also this capacity share has to be included among the chromosome information of the generic solution. An example of the solution coding referred to a single BESS is reported in Figure 1. The whole chromosome of a generic individual is obtained by repeating this schema for a prefixed maximum number of BESS (N_{BESS}).

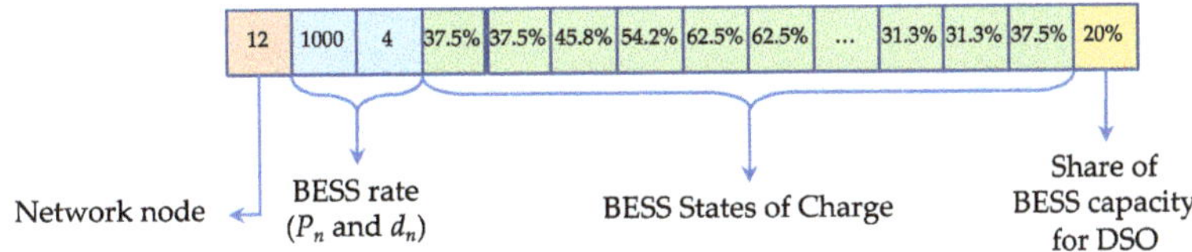

Figure 1. Chromosome section for a storage device.

In summary, the genetic parameters optimized simultaneously by the NSGA-II are:

1. **BESS position in the distribution network**

It corresponds to any MV network node, excluding the primary substation busbars because in those locations the BESS may affect only the loading of the HV/MV transformers, but it cannot provide support to local network contingencies. For real network studies, it is easy to exclude also additional MV nodes due to any kind of constraint the DSO planner wants to consider. The value of the specific gene is an integer number in the interval [1, N_{node}], where N_{node} is the total number of available MV nodes.

2. **BESS rate**

It is defined by two genes: one for the nominal power (P_n) and one for the nominal duration (d_n), so identifying by their product the nominal BESS capacity ($C_n = P_n \cdot d_n$). Nominal power and duration are the integer number within a minimum and maximum values ($[P_{n\_min}, P_{n\_max}]$ and $[d_{n\_min}, d_{n\_max}]$), given among the input data of the problem. Generally, d_{n_min} is the elementary time step used to represent the load/generation daily patterns (1 h), P_{n_min} is the minimum power required by the national regulation to participate in the ancillary services market, and P_{n_max} is the maximum power permitted for the direct installation of BESS (or any other generation unit) on the MV electric distribution system.

3. BESS daily energy schedule

It is represented by the sequence of the State of Charge (SoC) in each interval of the day. By using a time step (Δt) of 1 h, there are 24 genes for each typical profile used to represent the yearly customer behavior. Therefore, if more than one typical day is used (e.g., two semester profiles, four seasonal profiles, or multiple profiles for seasons, working days and weekends), the sequence of 24 genes is repeated accordingly. The SoC values are real numbers within the interval [0%, 100%]. However, these percentages are not related to C_n, but to the remaining capacity curtailed of the share for DSO (C_{own_TSO} in Figure 2). This choice allows preserving the soundness of the chromosome representation for the offspring solutions, built with the application of the genetic crossover and mutation operators. Indeed, by so doing, the new sequences of SoC remain always within the operational limits, avoiding overlapping with the semi-bands of capacity reserved to DSO.

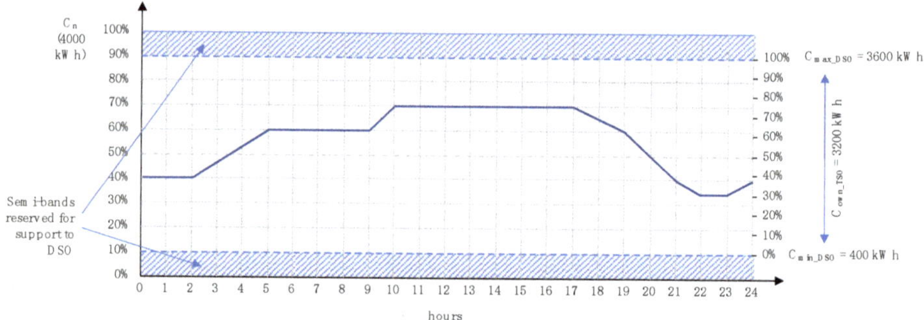

Figure 2. Decoding of the chromosome of Figure 1: BESS with P_n = 1 MW and d_n = 4 h; C_{DSO} = 800 kWh (20% of C_n) reserved for DSO; upper and lower semi-bands of 400 kWh; depending on the SoC, the virtual storage can be used as a BESS of 800 kW × 1 h or 400 kW × 1 h.

4. Share of BESS capacity for DSO

It establishes the rate of the "virtual storage" for DSO. From this value (C_{DSO}), expressed as a percentage of C_n, the upper and lower semi-bands of capacity are calculated ($C_{min\_DSO} = C_{DSO}/2$; $C_{max\_DSO} = C_n - C_{DSO}/2$) and always reserved to DSO in order to guarantee in any moment the availability of the virtual storage capacity in case of contingencies that could bring the distribution system to a critical emergency configuration (Figure 2). The residual capacity, $C_{own\_TSO} = C_n - C_{DSO}$, is managed by the private owner for energy trading and for offering ancillary services to TSO. The maximum available power of the virtual storage (P_{max_DSO}) may change hour by hour because it depends on the current SoC. If it is around half of C_n, it is reasonable the potential exploitation of the highest value ($P_{max\_DSO} = C_{DSO}/\Delta t$, with $P_{max\_DSO} \leq P_n$), while when the SoC is coincident to the lower or upper bound (C_{min_DSO} or C_{max_DSO}), the maximum available power takes its lowest value ($P_{max\_DSO} = C_{DSO}/2\Delta t$, with $P_{max\_DSO} \leq P_n$). Again, for this gene, the value is a real number within the interval [0%, 100%].

It is worth noticing that the decision variables (e.g., BESS SoC, share of C_n, rated power, and energy or duration) are continuous and, therefore, also the MO optimization solution space cannot be binary, as in the simplest version of the algorithm in the literature. For this reason, it has been decided to adopt a more efficient real codification. Indeed, this choice eliminates the accuracy problems related to the discretization of the solution space proper of the binary formulation that may not result adequately precise [27]. Furthermore, the formulation of genetic algorithms with the real instead of the binary coding is well suited for dealing with the gradual trend exhibited by continuous variable functions (i.e., small variations in the variables slightly alter the function values). Its implementation has required some adaptations to the typical operators of genetic algorithms (i.e., crossover and mutation), as detailed in [8].

3. Objective Functions

The advent of flexible resources in the distribution system during the deregulated power system era introduces new challenges to be faced, among which the correct cooperation between TSO and DSO is a cornerstone. Indeed, flexibility is essential for preserving the secure operation of the power system, but it is becoming central also for development and management of upcoming smart distribution networks. Moreover, the uncoordinated exploitation of flexibility from DER for TSO requirements may cause severe technical challenges to the distribution system, not originally planned for this scope.

Diverse TSO/DSO cooperation models are possible and have been analyzed by several scientific working groups [28,29], like various services market models as well [6]. In the paper, it has been assumed the opening of the ancillary services market (ASM) to the BESS located in the distribution system for the provision of primary and secondary frequency reserve. No local ASM has been supposed on the distribution level, but the DSO may request to reserve part of the BESS capacity for facing local contingencies. Two stakeholders have been considered with their goals:

- the BESS owners, whose goal is to maximize the energy trading revenues providing services to the TSO, and
- the DSO, that must operate its network energy-efficiently and cost-effectively within the technical constraints.

In the MO formulation, the DSO's objective function (OF) has been defined as the annual risk to violate any technical constraints (to be minimized), whereas for the BESS owners the OF is the cost/benefit ratio (to be minimized as well).

By optimizing the two OFs, it is possible also to analyze the limitation on the availability of the flexibility services for the TSO caused by the bottlenecks in distribution network. Indeed, if the location of a particular BESS is convenient for the distribution system, the DSO will reserve part of the available flexibility to resolve local contingencies, so reducing the amount of flexibility accessible to TSO.

3.1. DSO Objective Function—Risk of Technical Constraints Violation

Modern planning tools for designing the upcoming smart distribution networks need to overcome the traditional and extra conservative "fit and forget" approach, built around the fulfilment of the worst-case scenario, to move towards a risk-based procedure, that can correctly correlate the effectiveness of planning choices to the probability and the seriousness of possible contingencies [2].

An explicit and detailed assessment of the annual risk to violate technical constraints in a given distribution network requires probabilistic load flow (PLF) calculations, solved for each time step of the typical days used to represent the yearly behavior of distribution network customers [30]. The load/generation behavior in each hour has been assumed normally distributed. The risk assessment procedure starts with the definition of the impedance matrix $[Z]^b$, relative to the bth network configuration in the N-1 security analysis (where $b = 0$ means the network in normal configuration and $b > 0$ means the network reconfigured without the bth element), and the acquisition of the nodal current matrix $[I_{node}]^f$ in the fth hour of the typical daily profile. The results of the PLF are the nodal voltage $[V_{node}]^f$ and the branch current $[I_{branch}]^f$ matrixes (expressed in terms of mean value, μ, and standard deviation, σ, of a normal distribution), by which the probability (p_{tcv}) to overcome the voltage regulation band or the conductor thermal limit is estimated. Only the Nc operating conditions with non-negligible probability to violate the technical constraints ($p_{tcv} > 0$) are stored (Figure 3—case A), while the cases on which the extremes values of $[V_{node}]^f$ and $[I_{branch}]^f$ (assumed equal to $\mu \pm 3\sigma$) do not exceed the technical limits (minimum and maximum nodal voltages, V_{lim_min} and V_{lim_max}, and maximum branch current, I_{lim_max}) are disregarded (Figure 3—case B).

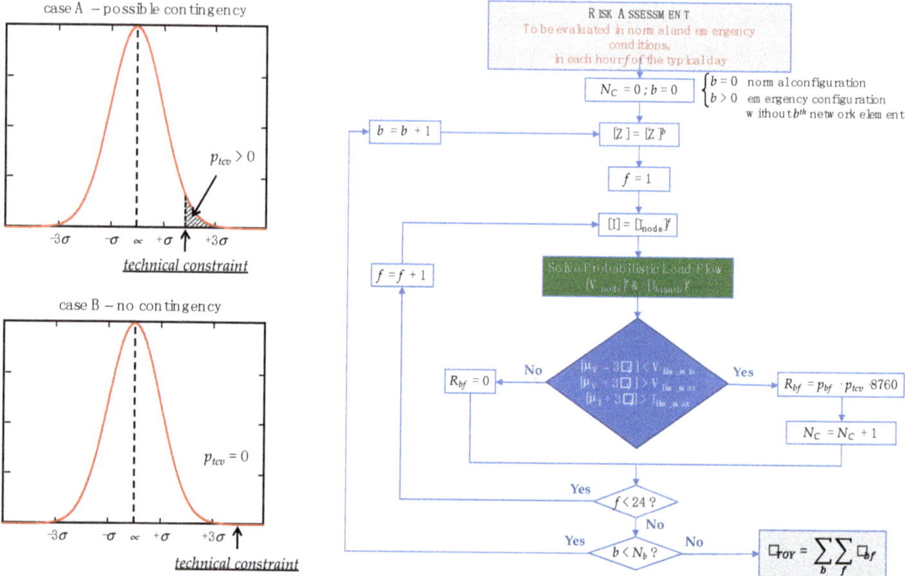

Figure 3. Identification of potential contingencies (p_{tcv}) and procedure for the total risk assessment.

In order to calculate the risk of technical constraints violation (R_{bf}) when the bth network configuration is in force during the fth hour of the dth typical day, the p_{tcv} is multiplied by the occurrence probability (p_{bf}) of the relative operating conditions. Such probability can be determined by simply multiplying the forced outage rate of the bth network element (FOR_b) and the occurrence probability of the specific customers' operating conditions (p_{fd}), because these two probabilities can be considered independent:

$$FOR_b = \frac{MTTR_b}{MTTF_b + MTTR_b} \qquad p_{fd} = \frac{nh_{fd}}{8760}, \tag{2}$$

where:

- $MTTR_b$ is the "Mean Time To Repair" of the bth network element, indicated in the paper with the symbol τ_b and assumed equal to 5 h for an overhead line and 8 h for a buried cable;

- $MTTF_b$ is the "Mean Time To Failure" of the bth network element, expressed by definition as $8760/\lambda_b$, where λ_b is the fault rate assumed in the paper equal to 0.12 [faults/(year·km)] for overhead lines and 0.03 [faults/(year·km)] for buried cables;
- nh_{fd} is the yearly number of occurrences of the fth hour of the dth typical day (i.e., the specific conditions of power injected or absorbed in each node by every customer). If a single daily profile is used to describe the customers' behavior in the whole year, then nh_{fd} = 365 h and the occurrence probability p_{fd} = 1/24; if two semesters are simulated, p_{fd} = 1/48, and so on.

For a distribution network, it is evident that $MTTF \gg MTTR$ (years compared to few hours). Consequently, $MTTR$ can be disregarded in the denominator of the first of Equation (2), and it is acceptable to assess the occurrence probability p_{bf} with the following approximated formula:

$$p_{bf} = \left(\tau_b \cdot \frac{\lambda_b}{8760} \right) \cdot \left(\frac{nh_{fd}}{8760} \right). \tag{3}$$

When the normal configuration is examined (b = 0), FOR_b is assumed equal to 1 and $p_{bf} = p_{fd}$. Finally, the risk component R_{bf} is expressed in hours of violation per year:

$$R_{bf} = p_{bf} \cdot p_{tcv} \cdot 8760 \left[\frac{\text{hours}}{\text{year}} \right]. \tag{4}$$

The sum of all these risk components, determined for each configuration in each interval of the typical days, represents the total risk (R_{TOT}) that characterizes the distribution network examined, i.e., the number of hours per year when it is possible to overcome a technical constraint.

When the total risk is greater than the acceptable limit fixed by the DSO planner, R_A, planning solutions have to be put in place to reduce R_{TOT} below R_A and make the distribution network robust enough for the whole planning period considered. Obviously, this system development has to be optimized by comparing the conventional network reinforcement with the exploitation of flexibility services from DERs, not only in terms of costs but also of residual risk. In the paper, only the resort to services from BESS has been considered, by exploiting when necessary the share of storage capacity ($C_{DSO\%}$) reserved to DSO. A simple linear programming optimal power flow (OPF), capable of finding the optimal scheduling of BESS, is used for tackling the particular contingency and nullifying the residual risk [3]. If the BESS location and/or the share of capacity for DSO are ineffectual for solving the contingency, the existing risk component will be reduced according to the flexibility resource availability.

3.2. BESS Owners' Objective Function—Services for TSO

The recent trend in opening the ancillary services market to resources located in the distribution system is creating new business opportunities for private investors. Indeed, the large diffusion of small non-dispatchable renewable generation units and the concurrent decommissioning of traditional fossil-fuelled plants are increasing the need for additional flexible and fully controllable resources, like BESS. Therefore, the introduction of new actors in the power system is expected, whose goal is to make a profit with storage devices mainly by providing ancillary services to power system operators. Obviously, the participation to the energy market is available as well, and it has been included in the following cost-benefit analysis (CBA), even if its weight is lower than the income from ancillary services provision.

In the paper, with the local ASM for the distribution system not yet available, the ancillary services provision has been assumed devoted only to TSO, even if DSO may limit this operating mode by acquiring part of the BESS capacity. Thus, the BESS owners' sources of income considered in the calculations are related to the following three applications:

- Arbitrage—it refers to energy trading strategies within an electricity market, aiming at purchasing energy from the grid when the price is low and selling it back to the grid at a meaningfully higher price (B_{arb}). It should be noted that this application, together with the influence on the distribution network operating conditions (nodal voltages and line currents), guides the optimization.
- Frequency containment reserve (FCR)—it is used for the purpose of primary frequency control. It must be fully provided within 30 s from the activation, remaining active up to 15 min. In the ASM, it is monetized with capacity payment based on availability and activation payment based on activated net energy (B_{FCR}).
- Automatic frequency restoration reserve (aFRR)—it is devoted to the secondary frequency control for restoring the frequency to its setpoint value and freeing the capacity needed by the primary control. It is characterized by an activation time between 30 s and 15 min, and its full capacity has to be guaranteed for more than one hour (variable from country to country). It is monetized with both capacity and activation payments (B_{aFRR}).

These BESS applications are mutually dependent, besides limited by the DSO reserve capacity, because they may be provided simultaneously, and their value relies on the active power exchanged with the grid. By optimizing the BESS daily scheduling of the stored energy, the multi-objective procedure adopted in the paper finds the optimal share of the BESS active power between arbitrage (and the influence on the distribution network operating conditions) and TSO services. For the further partition between the two frequency ancillary services, the heuristic choice of equally dividing the residual active power was taken.

By assessing the incomes from these three applications, in terms of net present values (NVP) within the whole planning period, and the storage investment (C_{BESS}), the BESS owners' point of view has been expressed as a cost-benefit ratio:

$$OF_{BESS\_owners} = \frac{C_{BESS}}{\left(B_{arb} + B_{FCR} + B_{aFRR}\right)}. \tag{5}$$

In the following, more details on the determination of cost and benefits are provided.

3.2.1. BESS Cost

A BESS is basically constituted by two components, the electrochemical battery and the system devoted to the power conversion. Thus, its installation cost can be split on a term proportional to the nominal power and another term relevant to the rated capacity. Regarding the share of BESS capacity reserved to DSO, several market mechanisms for remunerating this service are possible. In the paper, the simplest method has been supposed: the DSO refunds only the quote of the reserved share capacity. Regarding the cost term on power, proportional to the specific cost c_p (in €/kW), it is entirely charged to the private investor because the nominal power is entirely free until the SoC does not overlap the reserved upper and lower semi-bands (Figure 2). Consequently, the initial investment (I_{BESS}) for the private investor is defined as in (5):

$$I_{BESS} = \left(c_p \cdot P_n\right) + \left[c_e \cdot \left(C_n - C_{DSO}\right)\right], \tag{6}$$

where c_e is the unitary cost of capacity (expressed in €/kWh). c_p and c_e values depend on the BESS technology considered. The cost for the maintenance was assumed negligible in this paper even though it can be included in the model. If the BESS lifespan (N_{BESS_life}) is longer than the years' number of the planning period (N_{years}), the residual value (R_{BESS}) has to be considered and subtracted to the initial investment for a correct CBA:

$$C_{BESS} = I_{BESS} - R_{BESS} = \left(c_p \cdot P_n\right) + \left[c_e \cdot \left(C_n - C_{DSO}\right)\right] \cdot \left(1 - a^{N_{years}} \cdot \frac{N_{BESS_{life}} - N_{years}}{N_{BESS_{life}}}\right), \tag{7}$$

where $a = 1/(1 + d)$ is the actualization factor used to assess the net present value (NPV) of R_{BESS}, derived by assuming a constant discount rate (d) during the whole planning period. On the contrary, if $N_{BESS\_life} < N_{years}$, the investment on a new storage device has to be included.

3.2.2. Arbitrage

As aforementioned, arbitrage takes advantage of energy market price spreads (between off-peak and peak demand hours) that can produce value, even considering the efficiency of the BESS.

For each configuration of BESS proposed by the NSGA-II, the arbitrage term is calculated by subtracting the cost of storing energy in the device (charging phase) from the incomes of providing energy to the system (discharging phases), properly considering the round-trip efficiency of the BEES (i.e., the energy lost during the storage operation).

3.2.3. FCR Service

The FCR term is assessed by calculating the upward and downward power that the BESS can provide in each interval, depending on the scheduled daily profile of the energy stored. As assumed in the paper, in planning studies, the minimum time-step used to represent the customer variability within the typical day is one hour. Consequently, the offered FCR service is assumed constant within the same hour.

The power exchanged with the grid in the generic f^{th} hour is assessed as the difference between the starting and the ending state of charge (SoC):

$$P_f^{BESS} = \frac{SoC_f - SoC_{f+1}}{\Delta t}. \tag{8}$$

If the BESS is charging ($SoC_{f+1} > SoC_f$ being SoC_f the SoC at the beginning of the fth hour), it absorbs power from the distribution system, operating as a load ($P_f^{BESS} < 0$). If, on the contrary, the BESS is discharging ($SoC_{f+1} < SoC_f$), it delivers power to the distribution system, operating as a generator ($P_f^{BESS} > 0$). Thus, since ancillary frequency services have to be provided with a symmetric band, the width of the available semi-band in the fth hour (ΔP_f) is given by the minimum power that the storage device can provide, taking into account the BESS nominal power (P_n), the power injected to or absorbed from the grid in the fth hour (P_f^{BESS}), and the limits imposed by the DSO reserved capacity (C_{min_DSO}, C_{max_DSO}):

$$\Delta P_f = min\left\{ \left(P_n - \left|P_f^{BESS}\right|\right), \tfrac{1}{\Delta t}\cdot\left(C_{max\_DSO} - max\left\{SoC_f, SoC_{f+1}\right\}\right), \tfrac{1}{\Delta t}\cdot\left(min\left\{SoC_f, SoC_{f+1}\right\} - C_{min\_DSO}\right) \right\}. \tag{9}$$

In addition to the operating limits of the storage device, the availability of FCR can be limited hour by hour by the minimum amount of reserve (ΔP_{min}^{FCR}) that has to be provided in order to be eligible as the FCR service provider (it depends on the national regulation). It is predictable that this limit, lowered in the recent years due to the broad diffusion of renewables, will be further reduced (or even deleted) to allow the participation of many small DERs directly connected to the distribution system. If the semi-band calculated is lower than this limit, it is assumed that BESS cannot provide the FCR service for that specific hour.

It is worth noticing that the calculated semi-band ΔP_f is the whole amount of available power that the storage can offer in the fth hour for the considered TSO services. Thus, it can be split in accordance with the ratio r^{FCR} between the two frequency regulation services provided, or it can be dedicated exclusively to FCR ($r^{FCR} = 1$) or aFRR ($r^{FCR} = 0$).

The income from the FCR service is usually remunerated in capacity and energy. The first term is simply obtained by multiplying the cumulative amount of FCR available in the whole year by the average annual price of remuneration (p_C^{FCR}). The second term is estimated by assuming reasonable hypotheses on the average request of primary frequency regulation, based on historical measurements. Thus, the actual amount of energy used for this ancillary service in the fth hour is calculated by multiplying the available energy ($r^{FCR} \cdot \Delta P_f \cdot \Delta t$) by a heuristic factor f_d^{FCR} (assumed constant). Then, the remuneration of this energy is obtained by multiplying it for the corresponding hourly energy market price (p_{Ef}). By so doing, the total income from the FCR service in a generic year within the planning horizon is:

$$B_{yearly}^{FCR} = 365 \cdot \sum_{f=1}^{N_f} \left[\left(r^{FCR} \cdot \Delta P_f \cdot \Delta t \right) \cdot \left(p_C^{FCR} + f_d^{FCR} \cdot p_{Ef} \right) \right]. \tag{10}$$

The NPV of this frequency service in the whole planning period is finally derived as:

$$B_{FCR} = \sum_{i=1}^{N_{years}} a^i \cdot B_{yearly}^{FCR} = B_{yerly}^{FCR} \cdot a \cdot \frac{1 + a^{N_{years}}}{1 + a}. \tag{11}$$

3.2.4. aFRR Service

Operating reserves of this category are typically activated centrally with an activation time between 30 s up to 15 min. Differently from FCR, the aFRR may last more than one hour, but its maximum requested duration (Δt^{aFRR}) differs country by country (e.g., 2 h are requested in Italy).

The available semi-band offered for the aFRR service is calculated as $(1-r^{FCR}) \Delta P_f$. However, an additional constraint has to be considered, because the power offered must be provided constantly and continuously at least for Δt^{aFRR}. Consequently, the available semi-band for the aFRR service is obtained as the minimum of the following values:

$$\Delta P_f^{aFRR} = min \left\{ (1 - r^{FCR}) \cdot \Delta P_f, \frac{C_{max\_DSO} - max\left\{ SoC_f, SoC_{f+1} \right\}}{\Delta t^{aFRR}}, \frac{min\left\{ SoC_f, SoC_{f+1} \right\} - C_{min\_DSO}}{\Delta t^{aFRR}} \right\}. \tag{12}$$

The equations used to monetize this service are formally the same used for the FCR service, but with a different amount of maximum semi-band available, different capacity price, p_C^{aFRR}, and different heuristic factor, f_d^{aFRR}, used to estimate the average amount of energy provided.

3.2.5. Reduction Estimation of Ancillary Services for TSO

The exact amount of ancillary services provided to TSO and, consequently, the relative amount of incomes for the BESS owners may appear not directly dependent on the location of the BESS in the distribution network. However, being the storage devices connected to a given distribution network, their operation can be limited by the risk of violation of distribution system technical constraints. Indeed, if all these resources contribute simultaneously to an FCR request, they cause high momentary power flows that may exceed the maximum allowed overcurrent or cause excessive voltage deviations. The provision of secondary frequency control can also originate the same technical issues, even exacerbated because both services can be provided simultaneously. From this point of view, these services can depend indirectly on the position of BESS in the distribution network.

In order to assess these network limitations, the following procedure has been implemented. In normal operating conditions (Figure 4), for each hour of the typical day, a first probabilistic load flow (PLF) is executed without considering the provision of TSO ancillary services (but with the BESS typical scheduling). If the risk of technical constraints violation is greater than the acceptable one ($R_{TOT} > R_A$), the TSO ancillary services are suspended, and the virtual battery reserved to DSO is systematically used to solve or limit the network contingencies. In the other case, two additional PLFs are solved, by assuming that all the BESS installed are generating/absorbing their nominal powers simultaneously. If $R_{TOT} \leq R_A$, the TSO ancillary services can be freely provided, limited only by the respect of the DSO semi-bands and by the power exchanged for arbitrage (BESS scheduling). Otherwise, the TSO ancillary services are suspended again. To simplify the analysis, it has not been calculated the maximum amount of TSO services that can be provided without causing any network contingency. Still, the verification of technical constraints is always performed assuming the full provision of the available flexibility.

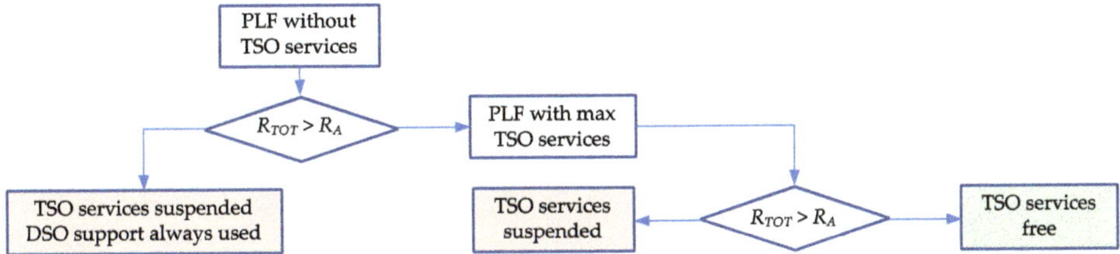

Figure 4. Identification of existing limitation in providing TSO services due to technical constraints violation in ordinary operating conditions.

In emergency configurations, the virtual battery is used to solve (or limit) possible contingencies during the repair of the faulted component, while the TSO ancillary services are not considered, because it is assumed that during emergency configurations of the distribution network, they are always suspended.

Once all these limitations have been identified, the total amount of flexibility for TSO services is estimated by summing in the typical day all the available regulation semi-bands ΔP_f. The result is finally compared with the same amount of flexibility calculated without considering the limitation introduced by the distribution network and the virtual battery of the DSO. In this way, it has been possible to estimate the reduction of the potential flexibility available for the TSO ancillary services, due to the DSO needs. Some examples of this procedure are depicted in Figure 5, that considers the storage device of Figure 2, with the assumption of $r^{FCR} = 1/3$ (i.e., one-third of the available band of flexibility in each hour is dedicated to FCR and two-third to aFRR) and $\Delta t^{aFRR} = 2$ h.

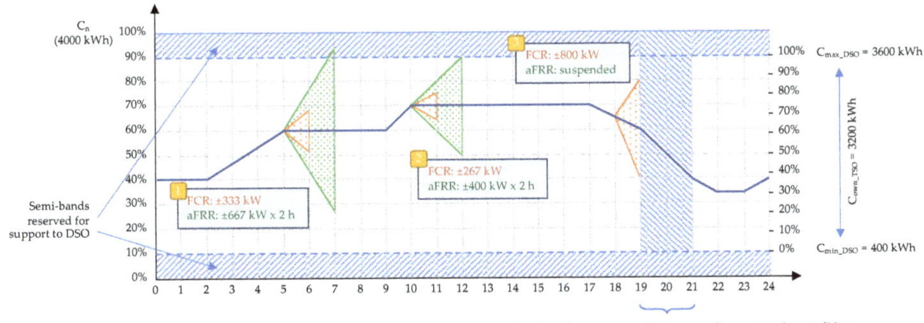

Figure 5. Examples of available flexibility for TSO services ($r^{FCR} = 1/3$ and $\Delta t^{aFRR} = 2$ h).

The first case of calculation refers to the preliminary estimation of the available flexibility for TSO without considering DSO limitation. Because the SoC in the fifth hour of the day remains constant (P_f^{BESS}

= 0), the semi-band coincides with the nominal power of the battery (P_n = 1000 kW). Thus, the flexibility for FCR is ±333 kW, and for aFRR is ±667 kW.

The second case (10th hour of the day) considers the presence of the virtual battery dedicated to the DSO. The SoC is still constant (P_f^{BESS} = 0). However, the proximity of the upper bound limits the available regulation semi-band to ΔP_f = (C_{max_DSO} − SoC_{10})/Δt = (3600 − 2800)/1 = 800 kW ($<P_n$). Thus, the flexibility for FCR is ±267 kW, while the flexibility for aFRR should be ±533 kW. However, because secondary control has to be guaranteed for 2 h, this flexibility is further reduced to ΔP_f^{aFRR} = (C_{max_DSO} − SoC_{10})/Δt^{aFRR} = (3600 − 2800)/2 = 400 kW.

The third case (18th hour of the day) is partially influenced by the presence of the two hours (19th and 20th) during which, in ordinary conditions, the TSO services are suspended due to technical constraints violations. Because the aFRR has to last for two hours, it cannot be provided from the 18th to the 20th hour. Instead, the FCR can be provided in the 18th hour, while it is suspended in the 19th and 20th hour. The available flexibility for FCR, in this case, is limited by the discharge of the battery (P_f^{BESS} = 200 kW), resulting in ±800 kW.

4. Case Study and Discussion

The proposed case study describes a real application of the approach for finding optimal BESS installation projects in a small network of the Italian distribution system (Figure 6).

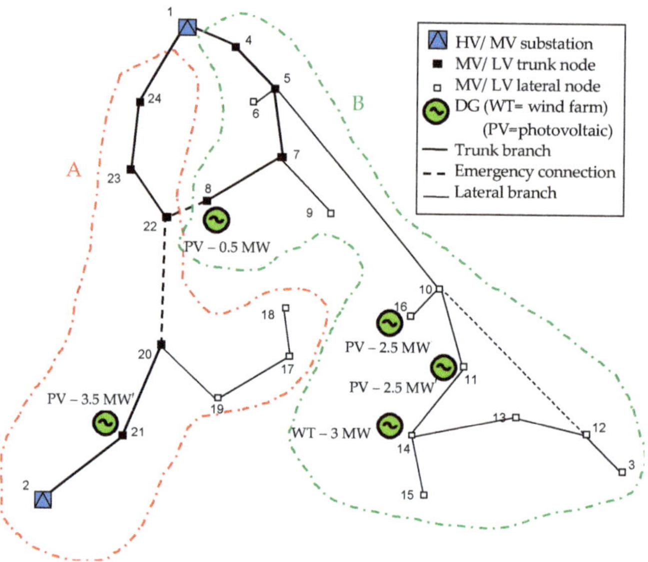

Figure 6. Test network.

Twenty-two MV nodes (9 trunk nodes and 13 lateral nodes), that supply energy to both MV and LV end-users (total rated power 13.8 MW), are fed by the bulk grid, via two primary substations, and by five RES generators: one 5 MW wind turbine (WT, in node 14) and four photovoltaic generators (PV, 0.5 ÷ 3.5 MW, in the nodes 8, 11, 16, 21). Loads and generators are modeled with typical daily curves. Two kinds of daily load profiles have been assumed for representing the residential customers (74%) and the agricultural ones (26%); the standard deviation adopted for the loads is equal to 0.05 pu. Figure 7 shows the used load patterns. Furthermore, the load and generation uncertainties are modeled with a normal probability distribution. The WT output power has been modeled with a constant mean value (0.50 pu), greater than zero, and a high and constant standard deviation (0.15 pu) during the day. While the PV generation, characterized by high production during the day and no output during the night, has been assumed with a standard deviation variable hourly (i.e., small at the sunrise and sunset, and significant, 0.03 pu, in the central hours of the day). The network is radially operated, but it is provided by some tie connections, usually open, that can be closed during emergency conditions. The test network may be considered located in a prevalently rural ambit since it is constituted by long overhead lines (fault rate equal to 12 fault/(year·100 km)) and the extended lateral branches. More details on the network characteristics are provided in the Appendix A. As a consequence of these characteristics, this network is electrically weak, and voltage regulation problems or overloads may occur due to the non-homotheticity between production and load demand.

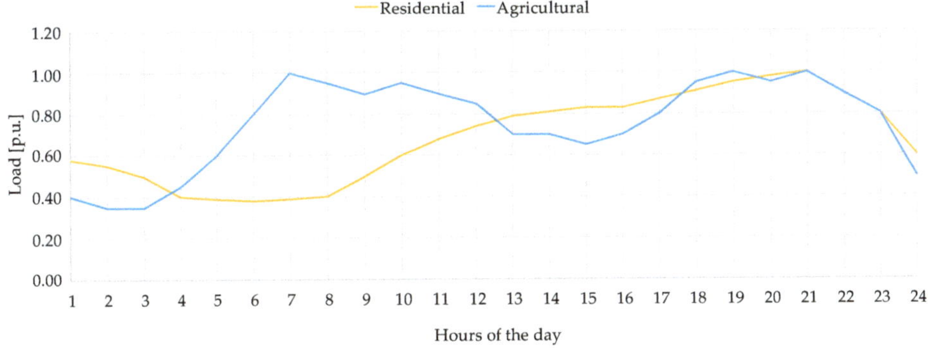

Figure 7. Case study: load profiles.

Table 1 reports the main parameters adopted in the case study, inherent to:

- the planning study (i.e., the period, the growth of the load demand, and the discount rate);
- the optimization algorithm (i.e., the population size, the maximum number of generations, and the maximum number of BESS considered for each possible configuration). Since the dimension of the optimization problem grows with the population size and the maximum number of generations, 500 individuals per population and a maximum number of 50 generations have been considered for limiting the computation burden;
- the BESS, regarding the technical characteristics (i.e., technology, range of rated power and duration), the economic aspects (i.e., power cost, energy cost, FCR capacity price, and aFRR capacity price) and some assumptions about the services provided (i.e., the share between FCR and aFRR, the minimum amount of reserve, the expected duration of the FCR, and the expected and the maximum duration of aFRR). The most commonly used Li-ion battery is the chosen BESS technology, and the relevant costs were assumed.

Regarding the main parameters of the optimization algorithm, a general rule often adopted for the genetic algorithm is that population size and number of generations have to increase with the dimension of the optimization problem (D_{OP}—number of variables to be optimized simultaneously). Considering that the solution coding adopted uses 28 genes for each BESS (Figure 1), and that the optimization has been limited to three storage devices, then $D_{OP} = 84$. From sensitivity analyses performed on the specific case study, population size and number of generations have been chosen equal respectively to 500 individuals and 50 iterations, representing a good compromise between quality of the results and calculation time. Indeed, it has been observed that the population size should be $5 \div 6$ times D_{OP} in order to achieve a high accuracy of the Pareto-optimal solutions set. Instead, the Pareto front does not improve significantly with the growth of generations after a minimum number of iterations ($0.5 \cdot D_{OP}$).

All nodes are eligible for BESS, but not more than 3 BESS are considered for each possible configuration. The threshold adopted for BESS owners CBA allows feasible solutions with a payback time comparable with the BESS lifespan (i.e., 12 years as in Table 1). Figure 8 shows the energy price daily pattern adopted for the monetization benefits.

Table 1. Main parameters used for the study.

Parameter	Value
Planning period (N_{years})	5 years
Yearly load growth	3%
Discount rate (d)	6.9%
Population size (genetic algorithm)	500
Number of generations (genetic algorithm)	50
Maximum number of BESS for each solution (N_{BESS})	3
BESS power cost (Li-Ion battery) (c_p)	300 €/kW
BESS energy cost (Li-Ion battery) (c_e)	200 €/kWh
BESS lifespan (Li-Ion battery) (N_{BESS_life})	12 years
BESS nominal power (range) ($P_{n\_min} \div P_{n\_max}$)	100 kW ÷ 3 MW
BESS nominal duration (range) ($d_{n\_min} \div d_{n\_max}$)	1 ÷ 8 h
Share FCR/aFRR (r^{FCR})	0.5
The minimum amount of reserve (ΔP^{FCR}_{min})	100 kW
Expected FCR duration (f_d^{FCR})	10 min
FCR capacity price (p_C^{FCR})	14 €/(MW h)
aFRR maximum duration (Δt^{aFRR})	2 h
Expected aFRR duration (f_d^{aFRR})	$3 \cdot f_d^{FCR}$
aFRR capacity price (p_C^{aFRR})	18 €/(MW·h)

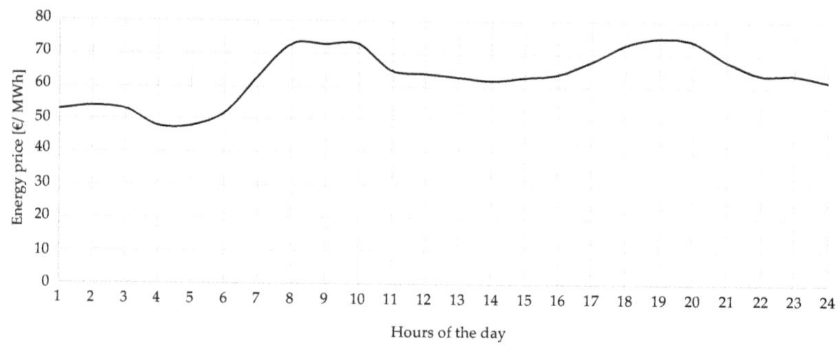

Figure 8. Case study: daily energy prices.

For validating the effectiveness of the procedure, the comparison between the optimal solutions and a base case, that does not use the support of the BESS, has been considered. In the base case (without any

BESS installed), network congestions occur during the evening (peak load, Figure 7) in the first branch of the B area (link 1–4, Figure 6), both in normal operations (network configuration showed in Figure 6) and in some emergency conditions. The total yearly overcurrent duration is equal to 735 h/year. For solving these contingencies, network upgrading would be necessary.

By applying the proposed optimization, the two OFs are calculated for each examined configuration. As a result of the optimization, the Pareto set is constituted by individuals that differ for the three BESS positions along the network, the nominal rates, and the daily schedules of charging/discharging cycle. The Pareto optimal front obtained for the proposed case study is shown in Figure 9.

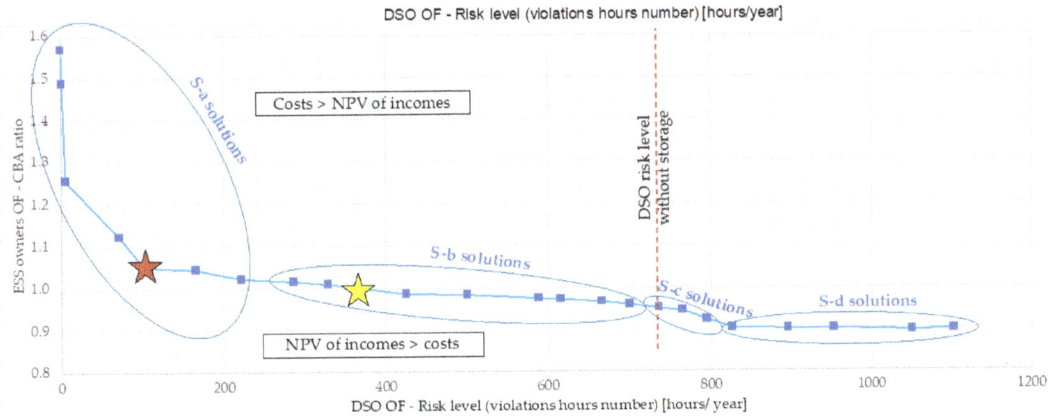

Figure 9. Case study: Pareto optimal front.

By analyzing the resulting optimal siting, a tendency of locating the BESS close to the 3.5 MW PV generator (Figure 6, A area) can be recognized. In that location, the BESS presence does not increase the risk of technical limit violations, due to the strength of the network (relatively short electric distance from the primary substations), in opposition to the majority of the nodes in the other part of the network (i.e., B area). Therefore, in the A area of the system, the BESS can be successfully exploited for providing TSO ancillary services (the frequency services), with negligible impact on the distribution network operation. For the BESS located in the B area, the suitable BESS reservation rate for the DSO use permits to provide ancillary services to the distribution network, with a valuable contribution to relieve the contingencies. Recurrent BESS size in the optimal solutions is in the range 0.1 ÷ 1.3 MW with duration 1 ÷ 7 h; the most recurrent size is 1 MW/2 MWh.

In the Pareto front, diverse types of solutions can be seen:

- solutions characterized by risk significantly smaller than the base case (S-a and S-b solutions, Figure 9), but several with a high CBA ratio (S-a solutions),
- profitable solutions for BESS owners without a reduction of the DSO risk for the base case (S-c solutions, Figure 9),
- most profitable solutions for BESS owners but with an increment of the existing DSO risk value (S-d solutions, Figure 9).

Starting from this classification, by analyzing each configuration, some comments arise:

- In the S-a solutions, some BESS mainly dedicated to distribution networks are located in the B area of the system; the other BESS are small in size and, in some cases, too small to offer the FCR/aFRR services in a profitable manner. The arbitrage service is not profitable due to the very narrow band in the daily energy prices during the day (Figure 8). For these reasons, the CBA ratio has the maximum value (1.569 vs. a risk equal to zero), in the S-a solutions. For this type of solution, the BESS in the A area can offer 100% of services to TSO, while for the BESS in the B area, the services are limited to the $72 \div 81\%$ of the total capacity. Globally, the overall solution can offer to the TSO a quantity in the range $83 \div 92\%$;

- The S-c solutions can be considered "neutral" from the DSO point of view because the BESS are located only in the A area that does not suffer for technical constraint violations (Figure 6), and they are dedicated to the FCR/aFRR services to maximize the BESS owners' profits (100% of their capacity for services to the TSO, with a residual risk at the same level, 735 h/year, of the no-BESS solution);
- The S-d solutions are close to the S-c in terms of the BESS owner profits but dramatically increase the DSO risk ($765 \div 1103$ h/year), due to a daily schedule that adds new violations in the network. Indeed, with respect of S-c solutions, in this case, some storage devices are located in the B area but without capacity reserved to the DSO (i.e., providing 100% of their capacity for services to the TSO);
- The S-b solutions are a compromise between the S-a and S-c solutions. The solution identified with a yellow star in Figure 9 can be considered as the "best compromise" between the different alternatives. In the selected solutions, the BESS owners can still obtain a profitable investment (even if quite small), while the DSO has a reduced risk in comparison with the base case (365 vs. 765 h/year). The new risk level is not the best for the DSO but, thanks to the BESS support, its network investments will be reduced. The goodness of the trade-off solution in the analyzed case study is limited by the BESS cost.

To investigate the impact of the BESS cost in the analysis, the CBA ratio in the Pareto front showed in Figure 9 has been calculated considering an expected 20% reduction of the BESS cost in the next years. It is important to remark that this sensitivity analysis does not modify the size and the type of solutions in the Pareto front but only the numerical values for the CBA ratio, because the BESS cost affects only the above mentioned OF. In the new conditions, the profitability threshold for the BESS owners (CBA = 1) is raised, and the new "compromise solution" can be identified with the red star in Figure 9, for which the total yearly overcurrent duration is equal to 102 h/year. In other words, following the expected reduction for the BESS cost in the next years, the proposed approach can identify a good compromise solution between BESS owners and DSO. For the DSO's point of view, it is essential to highlight that the BESS installation allows a partial risk reduction that can be completed by exploiting the flexibility from other local resources (active generators and loads). Moreover, even if not considered in the paper, BESS has also the potential to provide reactive support to the distribution network. Therefore, if local markets of ancillary services will be implemented, the profitability of BESS investment and the benefits for DSO operation can both get larger.

In Figure 10, the average daily schedules for the different BESS in the two areas of the network have been reported. In the A area, the BESS can be dedicated only to the owner profits because the DSO does not need support in this part of the network; the BESS change their SoC by maintaining an average energy level around 50%, to guarantee adequate support to the frequency service in upward e downward (TSO services). This assumption is confirmed by some sensitivity analyses for the frequency service remunerative prices; in particular, the simulation highlights that an increase in the remuneration (+15%) has similar effects of the BESS cost reduction, discussed above. On the contrary, the same reduction (−15%) in the remuneration for the frequency services jeopardizes the profitability of BESS investment because only the S-d solutions remain in a region where the NPV of the incomes are greater than the BESS cost (CBA ratio < 1).

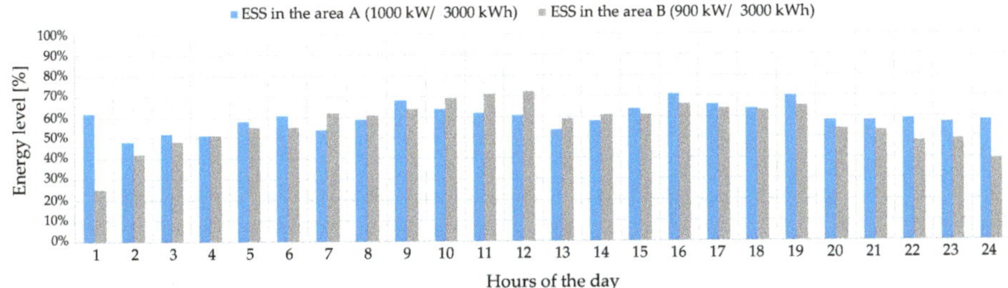

Figure 10. BESS charging/discharging profiles.

The BESS size, in the A area, is limited to avoid technical violations during the feasibility check for the frequency control participation. On the contrary, in the B area, the energy pattern mainly follows the DSO needs: indeed, the BESS discharge from the 19th hour to the 21st hour covers the peak load during the evening, when the power congestion occurs.

5. Conclusions and Further Works

BESS, among the distributed energy resources, can be considered the most flexible ones, and they can be suitably exploited for selling system services to the TSO and for solving temporary critical contingencies in distribution networks. The use of BESS will allow providing the services necessary for the management of RES in both distribution and transmission network, during the transition from the demonstration phase to the actual use of the flexibility product market. The paper presents an MO approach for optimizing the installation of BESS in distribution networks. The proposed process is a suitable instrument for the identification of the amount of flexibility (in terms of energy and power) that could be shared between DSO and TSO, without causing constraints violations on the distribution networks. The main novelty proposed in this paper is the optimization of both the objectives of maximizing the BESS owner profits and reducing the operation risk for the DSO. The proposed methodology is capable of producing a set of possible combination of BESS that are capable of offering valuable services to DSO for network operation while system services can be provided to the TSO. Further works will be devoted to the investigation of different remunerative schemes and/or regulatory frameworks regarding the DSO exploitation of the BESS owners. Because the formulation of the optimization problem is quite innovative and the results of different approaches are not available in the current scientific literature, in future research, the authors intend to apply different optimization techniques in order to identify which is more suitable for this kind of problem.

Author Contributions: Conceptualization, G.C. and F.P.; methodology, G.C., G.G.S., S.R.; software, G.G.S. and S.R.; validation, S.R., G.P. and F.P.; visualization, G.C.; data curation, S.R. and G.G.S.; writing—original draft preparation, S.R. and G.G.S.; writing—review and editing, F.P., G.C., G.P., S.R.; supervision, F.P.; funding acquisition, F.P. All authors have read and agreed to the published version of the manuscript.

Funding: This research has been supported by the project "Planning and flexible operation of micro-grids with generation, storage and demand control as a support to sustainable and efficient electrical power systems: regulatory aspects, modelling and experimental validation", funded by the Italian Ministry of Education, University and Research (MIUR) Progetti di Ricerca di Rilevante Interesse Nazionale (PRIN) Bando 2017—grant 2017K4JZEE, and by the project "BERLIN—Cost-effective rehabilitation of public buildings into smart and resilient nano-grids using storage", funded by the European Union under the ENI CBC Mediterranea Sea Basin Programme 2014–2020, priority B.4.3—grant A_B.4.3_0034.

Institutional Review Board Statement: Not applicable.

Informed Consent Statement: Not applicable.

Data Availability Statement: The network test data used in this study are presented in the Appendix A. **Acknowledgments:** The authors would like to thank the colleagues from EDF R&D (H. Baraffe, J. Fournel, G. Malarange, J. Morin) for the interesting discussion on this subject.

Conflicts of Interest: The authors declare no conflict of interests.

Appendix A

In the following the main parameters of the test network shown in Figure 6 are provided. Table A1 reports the main information of the lines (starting end and finish end, line length and line type) and the conductor parameters (cross section, resistance, reactance, capacitance, rated current). In Table A2 e in Table A3, respectively, the data about loads and generators are listed (node location, rated power, power factor (P.F.), and load/generator type). Loads are characterized by the daily load profiles shown in Figure 7. Generation is represented with representative production profiles according to the type of source.

Table A1. Main parameters of the line of the studied network.

ID	From	To	Length [m]	Cross Section [mm^2]	R [Ohm/km]	X [Ohm/km]	C [microF/km]	Rated Current [A]	Type
Br_1	1	4	864	95	0.320	0.125	0.350	200	BC
Br_2	4	5	1520	95	0.320	0.125	0.350	200	BC
Br_3	5	6	1003	95	0.320	0.125	0.350	200	BC
Br_4	5	7	2105	95	0.320	0.125	0.350	200	BC
Br_5	5	10	5000	95	0.320	0.125	0.350	200	BC
Br_6	7	8	2051	95	0.320	0.125	0.350	200	BC
Br_7	22	23	2398	95	0.320	0.125	0.350	200	BC
Br_8	23	24	1302	95	0.320	0.125	0.350	200	BC
Br_9	22	8	668	95	0.320	0.125	0.350	200	BC
Br_10	1	24	1846	95	0.320	0.125	0.350	200	BC
Br_11	7	9	1692	16	1.118	0.419	0.8	105	OHL
Br_12	10	11	1024	16	1.118	0.419	0.8	105	OHL
Br_13	10	16	1658	16	1.118	0.419	0.8	105	OHL
Br_14	11	14	2243	16	1.118	0.419	0.8	105	OHL
Br_15	14	15	2181	16	1.118	0.419	0.8	105	OHL
Br_16	14	13	2046	16	1.118	0.419	0.8	105	OHL
Br_17	13	12	1045	16	1.118	0.419	0.8	105	OHL
Br_18	3	12	3240	16	1.118	0.419	0.8	105	OHL
Br_19	10	12	2419	16	1.118	0.419	0.8	105	OHL
Br_20	17	18	804	16	1.118	0.419	0.8	105	OHL
Br_21	17	19	1257	16	1.118	0.419	0.8	105	OHL
Br_22	19	20	1325	16	1.118	0.419	0.8	105	OHL
Br_23	20	21	1860	35	0.520	0.430	0.9	190	OHL
Br_24	2	21	2788	35	0.520	0.430	0.9	190	OHL
Br_25	20	22	1215	35	0.520	0.430	0.9	190	OHL

BC: buried cable; OHL: overhead line.

Table A2. Loads main characteristics.

ID	Node	Rated Power [kW]	P.F.	Load Type
Load_1	3	100	0.900	AGR
Load_2	4	750	0.900	RES
Load_3	5	1800	0.900	RES
Load_4	6	520	0.900	RES
Load_5	7	1700	0.900	RES
Load_6	8	880	0.900	RES
Load_7	9	200	0.900	RES
Load_8	10	350	0.900	RES
Load_9	11	100	0.900	AGR
Load_10	12	200	0.900	AGR
Load_11	13	260	0.900	AGR
Load_12	14	200	0.900	AGR
Load_13	15	400	0.900	AGR
Load_14	16	430	0.900	AGR
Load_15	17	200	0.900	AGR
Load_16	18	550	0.900	AGR
Load_17	19	300	0.900	AGR
Load_18	20	70	0.900	AGR
Load_19	21	850	0.900	AGR
Load_20	22	550	0.900	RES
Load_21	23	1700	0.900	RES
Load_22	24	1700	0.900	RES

RES: residential; AGR: agriculture.

Table A3. Generators main characteristics.

ID	Node	Rated Power [KVA]	P.F.	Type
Gen_1	8	500	1.0	PV
Gen_2	11	2000	1.0	PV
Gen_3	14	5000	1.0	WIND
Gen_4	16	2500	1.0	PV
Gen_5	21	3500	1.0	PV

References

1. *The European Green Deal, COM (2019) 640 Final*; European Commission: Brussels, Belgium, 2019.
2. Abbey, C.; Baitch, A.; Bak-Jensen, B.; Carter-Brown, C.; Celli, G.; el Bakari, K.; Fan, M.; Georgilakis, P.; Hearne, T.; Jupe, S.; et al. *Planning and Optimisation Methods for Active Distribution Systems*; CIGRE: Paris, France, 2014.
3. Pilo, F.; Celli, G.; Ghiani, E.; Soma, G.G. New Electricity Distribution Network Planning Approaches for Integrating Renewable. *Wiley Interdiscip. Rev. Energy Environ.* **2013**, 2. [CrossRef]

4. Moreno, R.; Street, A.; Arroyo, J.M.; Mancarella, P. Planning low-carbon electricity systems under uncertainty considering operational flexibility and smart grid technologies. *Philos. Trans. R. Soc. Math. Phys. Eng. Sci.* **2017**, *375*, 20160305. [CrossRef] [PubMed]

5. *Directive of the European Parliament and of the Council on Common Rules for the Internal Market in Electricity, COM (2016) 864 Final*; European Commission: Brussels, Belgium, 2017.

6. "Smartnet Project". Available online: http://www.smartnet-project.eu/ (accessed on 14 November 2019).

7. Yoon, M.; Lee, J.; Song, S.; Yoo, Y.; Jang, G.; Jung, S.; Hwang, S. Utilisation of energy storage system for frequency regulation in large-scale transmission system. *Energies* **2019**, *12*, 3898. [CrossRef]

8. Celli, G.; Pilo, F.; Pisano, G.; Soma, G.G. Distribution energy storage investment prioritisation with a real coded multiobjective Genetic Algorithm. *Electr. Power Syst. Res.* **2018**, *163*, 154–163. [CrossRef]

9. Georgilakis, P.S.; Hatziargyriou, N.D. A review of power distribution planning in the modern power systems era: Models, methods and future research. *Electr. Power Syst. Res.* **2015**, *121*, 89–100. [CrossRef]

10. Saboori, H.; Hemmati, R.; Ghiasi, S.M.S.; Dehghan, S. Energy storage planning in electric power distribution networks—A state-of-the-art review. *Renew. Sustain. Energy Rev.* **2017**, *79*, 1108–1121. [CrossRef]

11. Bass, O.; Kothapalli, G.; Mahmoud, T.S.; Habibi, D. Overview of energy storage systems in distribution networks: Placement, sizing, operation, and power quality. *Renew. Sustain. Energy Rev.* **2018**, *91*, 1205–1230.

12. Sperstad, I.B.; Korpås, M. Energy storage scheduling in distribution systems considering wind and photovoltaic generation uncertainties. *Energies* **2019**, *12*, 1231. [CrossRef]

13. Sedghi, M.; Aliakbar-Golkar, M.; Haghifam, M.R. Distribution network expansion considering distributed generation and storage units using modified PSO algorithm. *Int. J. Electr. Power Energy Syst.* **2013**, *52*, 221–230. [CrossRef]

14. Zheng, Y.; Dong, Z.Y.; Luo, F.J.; Meng, K.; Qiu, J.; Wong, K.P. Optimal allocation of energy storage system for risk mitigation of DISCOs with high renewable penetrations. *IEEE Trans. Power Syst.* **2014**, *29*, 212–220. [CrossRef]

15. Ahmadian, A.; Sedghi, M.; Aliakbar-Golkar, M.; Elkamel, A.; Fowler, M. Optimal probabilistic based storage planning in tap-changer equipped distribution network including, capacitor banks and WDGs: A case study for Iran. *Energy* **2016**, *112*, 984–997. [CrossRef]

16. Hong, Y.-Y.; Lai, Y.-Z.; Chang, Y.-R.; Lee, Y.-D.; Lin, C.-H. Optimising energy storage capacity in islanded microgrids using immunity-based multiobjective planning. *Energies* **2018**, *11*, 585. [CrossRef]

17. Pombo, A.V.; Murta-Pina, J.; Pires, V.F. Multiobjective formulation of the integration of storage systems within distribution networks for improving reliability. *Electr. Power Syst. Res.* **2017**, *148*, 87–96. [CrossRef]

18. Babacan, O.; Torre, W.; Kleissi, J. Siting and sizing of distributed energy storage to mitigate voltage impact by solar in distribution systems. *Sol. Energy* **2017**, *146*, 199–208. [CrossRef]

19. Carpinelli, G.; Celli, G.; Mocci, S.; Mottola, F.; Pilo, F.; Proto, D. Optimal integration of distributed energy storage devices in smart grids. *IEEE Trans. Smart Grid* **2013**, *4*, 985–995. [CrossRef]

20. Wong, L.A.; Ramachandaramurthy, V.K.; Taylor, P.; Ekanayake, J.B.; Walker, S.L.; Padmanaban, S. Review on the optimal placement, sizing and control of an energy storage system in the distribution network. *J. Energy Storage* **2019**, *21*, 489–504. [CrossRef]

21. Xiao, H.; Pei, W.; Yang, Y.; Kong, L. Sizing of battery energy storage for micro-grid considering optimal operation management. In Proceedings of the 2014 International Conference on Power System Technology, Chengdu, China, 20–22 October 2014; pp. 3162–3169.

22. Riaz, S.; Mancarella, P. On feasibility and flexibility operating regions of virtual power plants and TSO/DSO interfaces. In *2019 IEEE Milan PowerTech*; IEEE: Milan, Italy, 2019; pp. 1–6. [CrossRef]

23. Gorostiza, F.S.; Gonzalez-Longatt, F. Optimised TSO–DSO interaction in unbalanced networks through frequency-responsive EV clusters in virtual power plants. *IET Gener. Transm. Distrib.* **2020**, *14*, 4908–4917. [CrossRef]

24. Motalleb, M.; Reihani, E.; Ghorbani, R. Optimal placement and sizing of the storage supporting transmission and distribution networks. *Renew. Energy* **2016**, *94*, 651–659. [CrossRef]

25. Coello, C.; van Veldhuizen, D.A.; Lamont, G.B. *Evolutionary Algorithms for Solving Multiobjective Problems*; Springer: New York, NY, USA, 2002.

26. Berry, M.; Cornforth, D.J.; Platt, G. An introduction to multiobjective optimisation methods for decentralised power planning. *Power Energy Soc. General Meet.* **2009**. [CrossRef]
27. Eshelman, L.J.; Schaffer, J.D. Real coded genetic algorithms and interval schemata. In *Foundation of Genetic Algorithms*; Whitley, L.D., Ed.; Morgan Kaufmann Publishers: San Mateo, CA, USA, 1993; Volume 2, pp. 187–202.
28. Celli, G.; Pilo, F.; Soma, G.G.; Canto, D.D.; Pasca, E.; Quadrelli, A. Benefit assessment of energy storage for distribution network voltage regulation. In *CIRED Workshop*; IET: Lisbon, Portugal, 2012.
29. Celli, G.; Mocci, S.; Pilo, F.; Allegranza, V.; Cicoria, R. An integrated tool for optimal active network planning. In Proceedings of the 17th International Conference on Electricity Distribution, Barcelona, Spain, 12–15 May 2003.
30. Celli, G.; Mocci, S.; Pilo, F.; Cicoria, R. Probabilistic optimisation of MV distribution network in presence of distributed generation. In Proceedings of the 14th PSCC, Sevilla, Spain, 24–28 June 2002.

Publisher's Note: MDPI stays neutral with regard to jurisdictional claims in published maps and institutional affiliations.

MDPI

St. Alban-Anlage 66

4052 Basel

Switzerland

Tel. +41 61 683 77 34

Fax +41 61 302 89 18

www.mdpi.com

Energies Editorial Office

E-mail: energies@mdpi.com

www.mdpi.com/journal/energies